# Apollo 11

## A Complete Guide to the Historic Moon Landing

*(The History of Nasa's Two Most Notorious Disasters)*

**Vivian Peters**

Published By **Simon Dough**

**Vivian Peters**



*Apollo 11: A Complete Guide to the Historic Moon Landing (The History of Nasa's Two Most Notorious Disasters)*

**ISBN 978-1-77485-529-4**

Legal & Disclaimer

The information contained in this ebook is not designed to replace or take the place of any form of medicine or professional medical advice. The information in this ebook has been provided for educational & entertainment purposes only.

The information contained in this book has been compiled from sources deemed reliable, and it is accurate to the best of the Author's knowledge; however, the Author cannot guarantee its accuracy and validity and cannot be held liable for any errors or omissions. Changes are periodically made to this book. You must consult your doctor or get professional medical advice before using any of the suggested remedies, techniques, or information in this book.

Upon using the information contained in this book, you agree to hold harmless the Author from and against any damages, costs, and expenses, including any legal fees potentially resulting from the application of any of the information provided by this guide. This disclaimer applies to any damages or injury caused by the use and application, whether directly or indirectly, of any advice or information presented, whether for breach of contract, tort, negligence, personal injury, criminal intent, or under any other cause of action.

You agree to accept all risks of using the information presented inside this book. You need to consult a professional medical practitioner in order to ensure you are both able and healthy enough to participate in this program.

# Table of Contents

# Introduction

Ranger 7

The Space Race is widely viewed with affection and reverence as an effort to reach the Moon which culminated in Apollo 11 "winning" the Race for the United States. It actually encompassed an entire spectrum of rivalries with both the Soviet Union and the United States which impacted the entire spectrum of technology from defense to launch satellites that would eventually be launched to Mars and orbit the other stars within the Solar System. Furthermore, the idea the idea that America "won" this Space Race at the end of the 1960s does not take into account how it was competitive Space

Race actually was in the launch of people into orbit and the significant impact to the space race. Space Race influenced in leading to the current International Space Station and continued space exploration.

The Apollo space program is among the most renowned and celebrated in American history, however the first time men were able to successfully land in space on Moon during Apollo 11 Moon at the time of Apollo 11 had complicated roots that go back more than 10 years, and was also the cause of one of NASA's biggest tragic
events. Launching astronauts onto the Moon was a dream in its own right however, the urgency of the government in creating Apollo's Apollo program was caused by it being a part of the Soviet Union, which spent in the 1950s largely abandoning its mark on the United States in its dust (and rocket fuel). In 1957, in a time where people were concerned about nuclear war and communism Many Americans were shocked to hear about it was the Soviet Union was successfully launching satellites into the orbit of.

One of the people who was concerned was President Dwight D. Eisenhower, who's space program was slowing a bit behind the Soviets in their space program. Between 1959 and 1963 between 1959 and 1963, in the years 1959-1963, United States worked toward putting astronauts and satellites into orbit through The Mercury program, however Eisenhower's administration was already preparing programs in the Apollo program in 1960, which was a year prior to when that first Russian satellite orbited Earth just two years prior to when John Glenn became the first American to orbit the Earth.

In May of 1961 the President John F. Kennedy addressed Congress and asked the country to "commit itself to achieving the objective, before the decade ends of putting a person in orbit on the Moon and safely returning him back to Earth." With the United States' lack of ability to even get one man on the Moon at the time, this seemed to be an unattainable goal and it's not even obvious that Kennedy believed that it was possible since Kennedy was reluctant to agree to NASA Director James E. Webb's initial requests for funding.

In during the decade of 1960, NASA spent tens of billions of dollars on missions to Moon which was the largest and most costly peaceful program of American history up to the point of completion. Apollo was made possible thanks to the tests carried out in the past by previous missions, including the legendary Ranger Program. The idea was originally conceived as the first step of the effort to put an astronaut in the Moon, Ranger was designed to capture images of the lunar surface to prepare for landings in the future. Prior to Ranger was launched, photographs of the Moon were only accessible via Earth-based telescopes. These did not provide the depth needed to identify safe locations for landing spacecraft. Ranger set out to fill this gap in understanding, as did many NASA missions in the 1960s, it exemplified both successes as well as the mistakes of the agency's initial period. But it was the Ranger Program would set important precedents, taking the first close-up images of the Moon and also landing the first American satellite on the moon's surface.

## Chapter 1: "Yes, To The Moon

At around four o'clock on Wednesday morning the 16th of July, 1969 at around 4 am, the Apollo 11 astronauts received their wake-up calls. Following their final physical examinations conducted performed by NASA astronauts and flight doctors, Neil Armstrong, Buzz Aldrin as well as Michael Collins had a final sitting-down session with director of Flight Crew Operations, Donald "Deke" Slayton.

In the morning, over a breakfast comprising steak, scrambled egg scrambles toast, orange juice and coffee Four men discussed the happenings of the next day. On the other side of the space, the artist Paul Calle sketched the astronauts as they ate. It was the first of many memorable photographs from that historic day.

Breakfast was served, and the group went into the room for suiting up to wear their outfits for the launch. It was the Apollo 11 crew had already gone through two weeks of virtual quarantine to reduce the possibility of exposure to disease before their space mission. After medical bio-sensory lead was placed on their torsos and legs, the three men changed into white nylon jumpsuits. With the help of skilled experts, they put on their space suits that were pressurized. Then, their helmets and their visors were closed and completely isolated from the world outside.

At 5:45 a.m. the astronauts left their spacecraft at the Kennedy Center for Manned Spacecraft Operations and hopped on an aircraft for transportation towards launch pad 39A. As they slung their portable fans with their hands, Armstrong, Aldrin, and Collins utilized the other hand to greet the journalists that were outside the building.

Although the majority of the world believed that they were brave explorers, the astronauts were extremely human. Neil Armstrong carried a comb as well as a bag of Lifesavers in his pocket and Buzz Aldrin pocketed pictures of his three children.

In Cape Canaveral, Florida, more than a half-million people been gathered to watch Apollo 11's launch. Apollo 11. The nearby highways were blocked, creating the creation of a 10-mile traffic jam. The flames were burning along the beaches as well as in the wooded areas around that space enthusiasts were camping all the night. Many vessels were moored on the adjacent Indian or Banana Rivers, set up for a bird's-eye perspective of this historic event. In the United States, over 25 million people were gathered around their TVs.

NASA has invited more than thousands of VIPs to attend the launch, which included around 250 lawmakers from Congress as well as half of governors of the states and the former president Lyndon B. Johnson, and the Vice-President of the moment, Spiro Agnew. Agnew. Jack Benny and Johnny Carson were among the famous faces from the world of entertainment. More than 3,000 journalists, from 56 nations, had been issued press credentials.

A viewing area was constructed three miles away of the landing pad including portable toilets, bleachers as well as water tanks and refreshment stations. Although it was early in the morning, Florida hot and humid conditions already took huge tolls on the audience. One of the most nervous of the spectators were the wives of astronauts, all who had travelled from their home within Houston and Cape Canaveral to witness the launch. Janet Armstrong's wife was Neil Armstrong. Janet was glued to the massive launch rocket from her yacht docked on the Banana River.

In the 1960s, demonstrations were commonplace. And on the hot summer day an estimated 150 protesters, led by civil rights activist the Reverend Ralph Abernathy, gathered within the vicinity of the landing pad. As the representatives of the Poor People's Campaign Abernathy as well as his supporters protested loudly against the government's "foolish use of funds which can be utilized to provide food to the hungry."

In the early morning the late Wernher von Braun the famous German rocket engineer who been in charge of designing the Saturn V rocket that would launch Apollo 11's Apollo 11 crew into space left his wife to kiss goodbye before departing at the Cocoa Beach Holiday Inn.

"Pray," he implored.

"Bye and good luck" she said.

To avoid the traffic jams, von Braun was flown by helicopter to Cape Canaveral. At the end of the counting down, von Braun who is known by the title of "Father of the American space program" turned his head in reverence and performed the Lord's Prayer.

As the astronauts arrived on The launchpad, the 363-foot-tall Saturn V rocket towered above with umbilical lines connecting it to a tower of support made of steel. After briefly looking up, Armstrong, Aldrin, and Collins rode into the elevator, where they found the space capsule atop the massive rocket.

Three-quarters of the way the elevator came to a stop and Buzz Aldrin was escorted onto a platform, was waiting, while his teammates were carried all their way up the highest point. Alone, and aloof from the world outside his spacesuit under pressurization Aldrin gazed up at the sun rising over the Atlantic Ocean. Aldrin, a veteran astronaut who had spent countless hours preparing for this landmark mission, was ready for a space flight for the second time in the space of five years.

In the capsule's boarding area which is known as the white room. Armstrong and Collins were the first to enter their way into the Apollo the Apollo capsule in first. Aldrin who had stood below to help ease the crowd before, was the last to board the tiny command service unit (CSM) that was to be the base for the astronauts for the remainder of eight days.

Fred Haise, the back-up lunar module pilot, just left the CSM and was christened Columbia after completing the 417 pre-flight test. Haise offered his fellow astronauts luck before taking his elevator to the ground.

The crew sat on their couches, which is NASA terminology for seats in the cockpit. The mission commander Armstrong was seated on the left, followed by Collins sitting on his right and Aldrin placed at the midpoint. As the capsule's hatch was closed the lead engineer on the launch pad Guenter Wendt handed Armstrong with a departure present--a crescent moon cut out of Styrofoam and then covered in

Tin foil. At the horizon's edge, unnoticed by anyone's naked eyes, was a small streak of moon, quarter of million miles away watched on Cape Canaveral.

After the launch pad's technicians closed the hatch of the capsule, Armstrong, Aldrin, and Collins could not have face-to-face interaction with any other human for over one week. In addition and tension, Apollo 11's Apollo 11 crew was keenly aware that any of a variety of catastrophic mishaps could leave them trapped in space for ever.

Pilot of the Command Module Michael Collins silently reflected on his upcoming role in the history of mankind: "Here I am, an unidentified white male, aged 30, 5'11-inches tall and weighing 165lbs and earning $17,000 per year living in the Texas suburb, sporting an unmarked black spot on my roses, mood shaky in the process of being shot off to the Moon, yes that's"the Moon."

## Chapter 2: Luna

The fascination of mankind with the Moon is a long-standing fascination. Astronomers have long been fascinated by its cyclical cycles and beguiled by the wonder of lunar and solar eclipses. For a long time, the source and structure of the celestial object were an unanswered question. The Romans identified it as Luna and the Greeks named it Selene as well as Artemis. In the emergence of Western civilizations it was known as the Moon, a term of Germanic origin.

in 1609 Italian Astronomer and physicist as well as mathematician Galileo Galilei began studying the Moon carefully, using the telescope, a device created a year earlier within the Netherlands. Utilizing the technique of "spyglass" as well as his artistic abilities, Galileo drew detailed sketches in his journal, Siderus Nuncius. At first, he believed that it was a smooth spheres, his drawings revealed the actual contours of the lunar surface, with craters and mountains surrounded by rugged terrain.

The following year, German mathematician and astronomer John Kepler extended Galileo's vision to anticipate the possibility the future of the exploration of space: "Let us create vessels and sails that can be adjusted to the heavenly ether There will surely be many people willing to take on the challenge of the empty space. While we wait, we'll create, for brave sky-walkers charts of celestial objects. I'll make it to honor the Moon and you Galileo and Jupiter."

The top echelons within the Roman Catholic Church soon branded Galileo as a heretic, for contesting the Aristotelianor Ptolemaic theory of there was a cyclic motion of the Sun, Moon, and planets revolved around an impermeable Earth and harshly condemned the Italian visionary's work, Dialogue on the Great World Systems. In retribution for his "blasphemous statements," Galileo was ordered to retract his observations and sentenced to serve the remainder of his life in confinement in an house located outside of Florence. Galileo was killed in 1642. However, it wasn't until the year 1992, that his pope John Paul II acknowledged that the astronomer of renown had been mistreated in the eyes of The Catholic Church.

As time passed and advances in science, man has gained an understanding of the relationship between Earth and the Moon, its sole natural satellite. The Moon. Moon is often referred to as a terrestrial planet because its dimensions and composition are like Earth, Mars, and Venus. The Moon however, isn't an actual planet but one of the satellites that orbit the Earth each 27.3 days. Since the Earth also orbits the Sun so the period between the new Moons is set by 29 days and 12 hours as well as 44 mins. The creators who created the calendar of today created moonths as a unit and later became called months.

The Moon's circumference is 2,160 miles (1/4th of Earth's) and its surface is just 1/13th the size. Although it is dwarfed by Earth but the Moon remains the second-largest moon in solar orbit and the biggest one in relation to the size of the planet it shares with.

The Moon revolves around Earth with an elliptical orbit rather that an equatorial planar. At its nearest orbital point (perigee) the distance from Earth towards the Moon is 221,463 miles and at the top the Moon's orbit (apogee) the distance is 252,710 miles.

The Moon is a synchronous spin which means that the amount of duration of its rotation about its own axis equals the amount of time required for a single circle within the Earth. Because of these two rotations, the exact face of the Moon is visible all the time in the sky. However because of small differences (librations) in the axis of its rotation, less than half Moon's surface Moon (59 per cent) is visible in one moment and the next. the remaining 41, which is referred to as the dark side of the Moon is never in sight of Earth. This is a misnomer as, even though it's being invisible from Earth the reverse of the Moon gets as much light as the side that is visible from Earth. The lunar days are longer than the days on Earth about 48 hours.

If the side facing the Moon is pointing towards the Sun it is visible when a full Moon can be observed. In New Moon phase and when it is turned towards the Sun in this case, the Moon is not visible to Earthbound observers.

Rarely, the straight alignment between three celestial bodies: the Sun, Earth and Moon result in a phenomenon referred to as an eclipse. If Earth is situated between the Sun and the Moon and casts its huge shadow on the Moon and a lunar eclipse can be seen through the night sky. Because the Moon's orbit tilted by about five degrees with regards to Earth's orbit lunar eclipses are not common during the full moon every night. Solar eclipses are an extremely rare event, the consequence of the Moon being directly ahead of the Sun and casting its shadow onto Earth.

As a pawn of Earth's gravity however, the Moon is able to exert an enormous influence over its planet. The lunar gravitational pull, which is to the side of Earth that is facing the Moon results in 2 large bulges (elevations in sea levels) that are continuously turning. The ocean's water continues to follow these bulges, creating Earth's low and high sea tides.

The lunar landscape is bleak and its conditions for living are brutal. The dark areas visible by the naked eye comprise the lunar plains, also known as maria (the Latin word for seas). Many experts believe that the lunar maria once held water. There are many reasons why the majority of the maria can be found in the vicinity of the moon. However, they make up just 16 percent of moon's surface.

The areas that appear lighter in the evening light of the moon are highland areas, or the terrae. Terrae range in height , ranging from tiny hills to mountains, and dominate the lunar topography.

Its lunar surface marked with craters resulting from the collision of comets and asteroids. More than 500 million of these craters cover the lunar topography. Some of them are only inches wide. they are all inches wide. The South Pole Aitken Basin, located on the other edge of the Moon is 2,250 kilometers wide and 12 kilometers deep. The Moon's surface is covered in regolith, which is composed consisting of fine dust, sand and other particles.

The surface temperatures of the Moon are extreme with an average of at 107 degrees Centigrade during the day , and the nighttime temperature is -153 degree Centigrade in the night. The

Moon's atmosphere is thin that it is virtually insignificant with only a few atoms of helium, argon methane, oxygen and carbon dioxide. This provides little protection from sun's damaging rays. Because light diffraction requires the presence of an atmosphere and atmosphere, the moon's sky is always a deep black. With the intense sunlight and deep shadows the lunar environment is not hospitable. With such harsh conditions, it's no surprise that there are no living things have been found within the Moon.

The origins of the Moon is the matter of intense controversy. Strict Creationists believe that the Genesis 1 Chapter of Genesis provides a clear explanation of the divine origins of the Moon. People who support the Fission theory believe that the Moon split off in a fragment from Earth due to powerful centrifugal forces. They believe that it left the basin a massive one that is now filled with the Pacific Ocean. Other researchers believe that the Moon originated elsewhere in our solar system but eventually attracted by Earth's gravitational pull - the Capture theory. The Co-formation hypothesis suggests it is believed that Earth as well as the Moon were created at the same time , from an accretion disk in the beginning. The most popular hypothesis in science is called the Giant Impact hypothesis - Theia A planetary body approximately as large as Mars was struck by Protoearth and then smashed away enough material to create the current Earth as well as the Moon.

Astronomers have speculated that the Earth was once home to two Moons that both formed at the time of the Giant Impact. The bigger of the two Moons three times larger in size and heavier by 25 percent than the other and thought to have drawn its smaller counterpart into orbit. the 5,000 mph collision of the smaller body against the larger caused the phenomenon known as the"big splat".

There is a strong evidence that supports the Co-formation as well as the Giant Impact hypothesis are both confirmed by geological analyses from Moon rock. A large portion of the moon rocks studied are believed as being 4.6 billion years old, which is the same amount of time as the oldest fossils.

Through the ages throughout history, throughout history, the Moon was used as a navigational and chronological marker. Even though it's not in brightness of Sun and reflecting only seven percent of its sun's light in the process, the Full Moon is still the brightest nighttime object in the sky. In the phase of crescent, it's only one-tenth the brightness of the full Moon.

In the event that the Moon is in the Horizon, it appears to be bigger however this is just an optical illusionsince it actually 1.5 percent smaller due to being further from the observer by that is up to one Earth radius. The Moon is at its highest in the winter months full Moons have been a source of light for numerous generations of night-time travellers.

Human fascination for the Moon as well as stars and planets developed into the romantic idea of space travel. In 1865, the novelist Jules Verne published From Earth to Moon A fictional tale about the lunar mission. In the story the rocket ship is launched from a huge cannon, which is called Columbiad. With an eerie prescience Verne's vehicle, manned, was launched from Florida then orbited the Moon before splashing into the Pacific Ocean.

Just over 100 years later, Verne's dream would be realized, and his method of travel was distinctly futuristic.

## Chapter 3: Vergeltungswaffe

Launching a vehicle into space requires an enormous thrust to counter gravity's force on Earth. The long-standing, but not well developed science of rocketry was found to be the sole method of producing this thrust.

From 1232 onwards as 1232, the Chinese employed rockets powered with gunpowder for fireworks during performances. in 1281 Italians who were from Bologna used rocket-propelled arrows in their battle against the state of Forli. adversaries from Forli The feared weapon a"rochetto," meaning "cylindrical thread."

The first rockets were powered by solid fuels, specifically gunpowder. Although solid fuels can possibly propel the rocket with sufficient speed to get to the outer space, a significant issue was that once the fuel was ignited it was impossible to control over the speed of combustion, or even the quantity of thrust.

born in Russia on the 17th of September 1857. Konstantin Edvardovich

Tsiolkovsky studied math, physics, and astronomy. He later utilized a large portion of his inventive energies to research in rocketry. Tsiolkovsky's work has led him to believe that a mixture of rocket fuel composed of liquid oxygen and hydrogen could generate a lot more force than black powder. In the process of mixing two liquids inside a sealed metal chamber, and then igniting them Tsiolkovsky believed that the gasses expanding generated by the explosion could be released through holes at high speeds, thereby propelling the rocket and their payload into the reverse direction. In the Russian scientists' Formula of Aviation defined the relationship between the speed and the mass of a rocket in relation to the propulsion fuel it uses. Tsiolkovsky determined that a speed at 18,000 miles/hour was required to overcome gravity force of the Earth. He was also able to determine that the most effective method to accomplish this was to employ the multi-stage launch rocket.

German mathematics professor Hermann Julius Oberth, born in Transylvania on the 25th of June 1894, wrote extensively regarding space travel in the 1923 work, The Rocket into

Interplanetary Space. A year later in his own book, Way to Space Travel, Oberth outlined the feasibility of making use of liquid-fueled rockets. The year, the same time period Oberth as well as other German rocketeers founded"VfR," which stands for the Verein fur Raumschiffahrt (VfR)--the "Society for Space Travel."

Robert Hutchings Goddard, born on the 5th of October, 1882. He is recognized as America's first rocket scientist. He was a native of Massachusetts, Goddard was educated at Worcester Polytechnic Institute, and later taught Physics in Clark University.

The New Englander's enthusiasm for rocketry started during his youth and ultimately was his career. At the age of 27, Goddard published A Method of Exploring Extreme Altitudes and proposed that the launch of a rocket from Earth could eventually reach the Moon. Like other scientists Goddard's time, the young rocketeer was confronted by many skeptical people. In the month of January, 1920 it was reported that the New York Times harshly criticized Goddard's assertion the possibility of using rockets to be employed to explore space: "He seems only to be ignorant of the theories that are taught every day in high school." 39 years later, when Apollo 11 raced to the Moon the renowned paper published a retraction of its piece of criticism Goddard.

Goddard created his first liquid-fueled rocket at the farm of his aunt at Auburn, Massachusetts in March of 1926. The rocket was dubbed Nell the 10-foot-tall, 10.25-pound rocket was powered by liquid oxygen and gasoline stored in fuel tanks that were connected via rigid tubes to an engine of a smaller size. When the gasoline and oxygen mixture was ignited in the combustion chamber the hot gases burst out through a tiny nozzle located in the middle of the rocket. As the rocket soared into the sky at 60 miles per hour the first rocket launched by Nell lasted just 2.5 seconds. It reached an altitude of just 41 feet before reaching in 184 feet below the target; nevertheless it was an important milestone in the field of rocketry science.

After meeting with a meteorologist from Clark University, Goddard determined that the climate of New Mexico was ideal for all-year-round rocket launches. In July of 1930 Goddard along with his wife and four of his assistants, along with a truck loaded with rocket equipment, were relocated to an area that was called Eden Valley, near Roswell, New Mexico. In the area, Goddard set up an experimental rocket science lab and testing range that included the launch pad as well as a tower.

Derisively known as "Moony" Goddard by his critics, the ambitious however, highly private rocketeer got no support from the government. In the span over four decades, the philanthropist Daniel Guggenheim provided Goddard with an annual $25,000.00 grant, and famed pilot Charles Lindbergh helped raise additional funds that allowed Goddard to pursue his goals.

With the passing of time Goddard's rockets got more sophisticated, with the addition of Gyroscopes. In 1929, Goddard launched the first instrument-containing rocket, which carried a thermometer, barometer, and camera high into the sky. Another of his rockets that was powered by liquid achieved sound speed (Mach 1,) in the year

1935. Goddard later developed an engine that could propel 1.5 miles high at speeds of 550 miles an hour.

Goddard continued to test rockets in his desert home for the remainder of his time. Despite his numerous accomplishments, Goddard was never able to attract in the U.S. military in rocket-propelled weapons. He eventually granted more than 200 patents Goddard kept pushing the boundaries of rocket science technologies up to his demise in 1945. In his final hours Goddard gave a glimpse into the future "It will be a matter of imagination of how much we can accomplish using rockets. I believe it's appropriate to say that there is nothing you've seen so far."

Following the path following the footsteps of Robert Goddard, Wernher von Braun eventually became the most popular rocket scientist in his time in the twenty-first century. Born on the 25th of March 1912, in Wilintz Germany von Braun developed a fascination with rocketry and space exploration from an early age, and was a fan of the science fiction novels of Jules Verne and H.G. Wells. After having read those futuristic stories, von Braun was "filled with a romantic lust," and "longed to take flight through the sky and discover the mysteries of the world." To enhance his knowledge of science, von Braun carefully studied the technical texts that were written by Herman Oberth.

In his early years, von Braun caught the interest of the locals by firing rockets through the bakery and an apple stand and his father later described the incident as one that was filled with "broken window" as well as "destroyed floral gardens." In one instance the young man sewed six large fireworks that he bought from the store to his pull-wagon made of wood. Following ignition, von Braun attempted to drive his rocket-powered vehicle along the sidewalk, and terrified pedestrians jumped away from the path. The police brought Wernher into custody for questioning following the erroneous experiment, but he was released to his father who promised to be responsible for the punishment he handed out to him. Despite his mishaps Von Braun's interest in science never waned, and when he was still in his teens He joined the club for rockets, Verein fur Raumschiffahrt.

At the age of 23 von Braun graduated from Friedrich-Wilhelm University in Berlin with an PhD in physics. The research topic for his dissertation was rockets that were liquid-fueled. In the beginning of his career, von Braun worked for the Society for Space Travel, alongside other rocket scientists who were a part of the same dream of space travel.

Five feet tall and eleven inches tall, von Braun was handsome and square-jawed with hair that was thick and blond hair. Athletic and proficient in many languages The astronaut was charming as well as friendly and had a wide range of interests that included music (he played the cello and piano) and religion, philosophy, geography and politics. Von Braun was also a talented writer, a spellbinding speaker, expert draftsman and pilot.

The Treaty of Versailles, which officially brought an end to World War I, punitively restricted German production of arms. The treaty, however did not contain any provisions regarding rockets that were at the time considered to be viable weapon of war. Therefore it was the German Army assigned artillery officer captain Walter Dornberger, the task of determining the best way to make use of this loophole.

In 1932, just a year prior to when Adolf Hitler rose to power, von Braun and his co-workers were hired by Dornberger to create rockets that could be used by the military. Von Braun later defended the choices made by his rocket team as steps towards their ultimate goal: "We were interested in only one thing, which was the study of the space."

After World War II, when Nazi atrocities were revealed in the public eye, von Braun was repeatedly asked to explain his first career path. "...We required money and the Army was willing to assist us. In 1932 the idea of war was a nonsense. In 1932, the Nazis were not in control. We had no moral scruples about the future potential usage of our invention."

As a military officer, von Braun began his period at Kummersdorf-West, which was an artillery test ground situated to the south of Berlin. After 1937 von Braun and his rocket-development team relocated to Peenemunde situated located on the Baltic coast, close to

Usedom. Von Braun was appointed the technical director of the Army Research Center, and was charged with the development of the first ballistic missiles in the world.

After arriving in Peenemunde, von Braun joined the Nazi Party, perhaps naively not realizing the consequences of his choice. Later on Von Braun claimed that the decision was "officially requested" to be a part of the fascist group If he had refused the invitation, it would have meant the end of "the activities of my life."

"My membership within the party didn't require any political involvement," von Braun explained.

The year was 1940. von Braun joined the notorious paramilitary Schutzstuffel (SS) at the death of its notorious chief Heinrich Himmler, and was given an rank called Untersturmfuhrer (Lieutenant). Through the rest period of his existence, von Braun would downplay his SS membership, noting that he didn't use his officer's rank in official correspondence. He also was in the black dress uniform, complete with armband with swastika only when absolutely required. As the war grew more intense and the Nazi bloodbath grew, von Braun was promoted to Hauptsturmfuhrer (Captain) and later Sturmbannfuhrer (Major).

The German Army was commanded by von Braun to develop operational ballistic missiles. The final product was Aggregat-4 (A-4) that the German Propaganda Ministry later renamed the Vergeltungswaffe-2 (V-2) which translates to the weapon for vengeance.

The V-2's winged rocket, measuring 46 feet long and weighing 14 tons was powered by ethyl Alcohol and liquid oxygen. It was anchored through four fins, and 4 rudders. Two gyroscopes that were mounted inside the nose, beneath the warhead that explodes, directed the weapon towards its destination. The missile was able to hit targets as far as 500 miles from the launch location. At 2,500 mph and equipped with a warhead of 2,200 pounds The V-2 was deadly and dangerous.

After two failures in tests between March and August of 1942 The first V-2 was launched successfully on the 3rd of

October the year. The test missile was able to reach 60 miles in altitude and left a lasting impression on the rocket scientist Krafft Erichke "It appeared like a flaming sword soaring through the sky...It is extremely difficult to explain what you feel when you are on the edge of a new era...We knew that the age of space was about to begin."

In the beginning of July 1943 von Braun and his military boss, Walter Dornberger, briefed Adolf Hitler on the V-2 rocket. They told Hitler Fuhrer about the rocket, which was armed with explosives that weighed 2,000 pounds could be fully capable of destroying London as well as the British could not take it down. In typical megalomaniac fashion, Hitler immediately ordered production of 2,000 V-2 missiles every month.

The first V-2 rockets were fired against London within a few days of D-Day (June 6 April 1944). Famous CBS News anchorman Walter Cronkite who, at the time, was the UPI war reporter, vividly described the terrifying attacks in describing the V-2 rockets in terms of "devilish arsenals."

Just five minutes after the launch, the terrifying rockets terrified Londoners who were used to hearing sirens of air raids as well as the sound from approaching German bombers before the imminent attack. Instead, the unaware civilians were awed by the sound of a "ball in the air" and an "terrible break." One person expressed the difficulty caused by the V-2 rockets: "There was no alert...We were not warned at all."

There were a total of 4,000 V-2-fired rockets were fired upon Allied target areas throughout England, France, and Belgium in the course of World War II; 1,403 of them struck London as well as various other locations in south England. A total of 5,400 people, nearly two-thirds of them civilians, were killed due to V-2 rockets.

Following Allied bombing attacks targeted Peenemunde the V-2 production facilities were relocated to an underground location within the Harz Mountains, near Nordhausen. Nearly five thousand workers involved during the building of Nordhausen were inmates from concentration camps.

Nordhausen facility were prisoners of concentration camps.

The supervisor for Nordhausen, the Nordhausen construction site, which was located in a former 35 million cubic-feet anhydrite mine was notoriously brutal SS General Hans Kammler. Prisoners, such as Russians, French, Poles and, later Jews were transferred from the Buchenwald concentration camp to Dora near Nordhausen. The prisoners were given the task of building caverns that were larger and were subjected to crowded horrible conditions. The damp, cold air, stale and dusty, with no proper ventilation dangerous for their breathing systems. There was neither running water nor sewage facilities and the inmates were made to sleep in bunk beds that were open, which were stacked four-high. Thus, the spread of dysentery, pneumonia and typhus were common.

A few of the more aggressive prisoners tried to disrupt the rockets in the process of assembly and those who were caught playing with missiles, or those who were deemed to be slackers were severely punished and/or killed by

Nazi supervisors. A total of 20,000 prisoners perished at Dora/Nordhausen. In final, far more died in the process of building V-2 missiles than were killed in the horrific weapons that were used during the this conflict.

Although von Braun denied playing any part in the choice to employ slave labor or the administration of the work details however, he was surely aware of the terrible conditions. Von Braun and his colleagues who were involved in the inhumane conditions at Nordhausen were repeatedly questioned in the following years. Von Braun biographer, Michael Neufield accurately described the involvement of the rocket scientist with the Nazis as a way to pursue his goals of space exploration as an "Faustian deal."

As an Nazi pawn von Braun never lost sight of his goal to utilize rockets for purposes other that weaponry. After he wrote his article in a journal of science advocating rockets as a means of international mail delivery Gestapo Director Heinrich Himmler ordered the rocketeer detained for "lack of dedication" to the military effort, and also for fabricated allegations which claimed that von Braun was associating with Communists. Walter Dornberger, von Braun's military supervisor, persuaded Hitler to free him from jail by arguing that his knowledge was vital for in the program V-2.

Von Braun and his colleagues were always dreaming that they could use rockets to achieve peaceful solutions. Following an early V-2 test launch, a German scientist was heard to say: "There goes the world's first spacecraft."

The spring of 1945 when the defeat of Germany appeared likely, von Braun gathered his team of rocket designers and engineers to discuss the direction they would take for their rockets. Conscient that the final battle was fast approaching the time to make a decision, he stated his alternatives: "We despise the French and are utterly terrified of the Soviets and we don't believe that the British could afford us, which means we are left with the Americans.

On May 2nd of the same date, von Braun and his colleagues surrendered to American military forces. They ignored an order that had been issued just six weeks prior by an elated Adolf Hitler, instructing the rocketeers to eliminate everything related to the V-2 program, including paperwork and equipment. to the V-2 programme, von Braun and his colleagues put nearly 14 tons documents within an old mining facility in Bleicherode Then, they dynamited the shaft in order to seal the contents. When they surrendered, the rocket group handed over to U.S. military officials some three hundred detailed reports as well as more than 500,000 blueprints for rockets kept in the secret location.

The V-2 facility in Nordhausen was located within the Soviet Union's postwar occupation zone. In the end, while recovering just 20 V-2 rockets, Soviets recruited their fair part of German experts in nuclear energy, aviation electronic, radar, as well as rocket sciences. Prior to the Soviet control of Nordhausen in the year 1945, Major James Pottamill, a ranking officer of the U.S. Army Ordnance Division was in charge of removing parts of nearly 100 V-2 rockets, as well as the removal of a "large collection of manuals, plans and other papers." The documents were shipped in United States. United States, before the Soviets discovered the cover-up.

Werner von Braun and his colleagues emigrated from Germany to United States beginning in September of 1945. The moment a tiny "advance security" of high-ranking rocket scientists crossed into the Saar river from Germany to France on their way to the plane that would carry these scientists to United States, von Braun warned his colleagues about how their decisions affected them: "Well take a good look around Germany and fellows. It's unlikely to be seen for some time."

By the middle of 1946 in the middle of 1946, around 118 German rocket researchers, technicians and their families had moved into their new homes in Fort Bliss, Texas, under the direct oversight from the U.S. Army's 9330th Ordnance Technical Service Unit. The huge effort to relocate them was initially dubbed Operation Overcast, but later changed to Operation Paperclip (in reference to the clasps made of metal used to tie the Germans to their papers of immigration). The hiring and relocation of German researchers and technicians was a good thing, considering that it was the case that the U.S. military had no missile technology, not even during the initial stages of planning which was as advanced similar to the V-2 missile.

Operation Paperclip, authorized by President Truman on September 6, 1946 was supervised through the Joint Intelligence Objectives Agency (JIOA) and was supervised by the Joint Chiefs of Staff (JCS). It was the first time that Germans had been officially designated to be "wards from the Army." Navy Captain Bosquet Wev director of the JIOA was required to the arm of American State Department officials, some of whom were agitated about admitting "ex-Nazis" in America. United States as "invited guests." Wev argued that the German scientists and technicians who were enlisted in the Soviet Union were a "far more serious security threat" in America than those who had "former Nazi affiliations" or present "Nazi allegiances."

The newly arrived Germans were located in a remote area in Fort Bliss, occupying dilapidated barracks, which were manned by a mess hall and a recreation club. Military personnel were accountable for the security and wellbeing of the Germans who had not yet issued passports. Separated from civilians The immigrants were allowed to leave the base without being accompanied by a guard. Most of the time, the sequestration was intended to protect the population because many civilians within the vicinity were extremely concerned about their new "ex-Nazi" neighbor.

Von Braun tried to ease the fears of Americans who were skeptical of relocating his rocket crew into his home country of the United States: "We are certain that the complete understanding of rocket science can alter the conditions of the world in the same way as the proficiency of aeronautics and this can be applied to the military and civilian aspects of rockets' use."

in 1947 von Braun was allowed to return to Germany to wed his 18-year-old cousin Maria von Quistor, with the agreement that the couple were to immediately return home to United States. On their honeymoon, the couple rented the house with American MPs they were assigned to prevent von Braun from being kidnapped by Soviet intelligence agents.

In 1949, German researchers and techs were loaded onto an auto and driven over the Mexican border in El Paso. The bus then was turned around and then returned to the border patrol station at El Paso, which is where the Germans were issued entry visas, which can later applied for American citizenship.

The surge of German creativity was not restricted to the rocket scientists of Fort Bliss. After the end in World War II, the U.S. military relocated nearly 1600 German engineers, scientists engineers, and techs to America.

The White Sands Proving Ground (an annexe of the army's Aberdeen Proving Grounds in Maryland) located 40 miles to the northeast from Fort Bliss, near Los Cruces, New Mexico, served as the launch point for the V-2 rockets that were captured. In the desert that was arid and isolated in Tularosa Basin, the Tularosa Basin, von Braun's rocket team merged with Project Hermes which was an unguided missile project that in the Army Ordnance Department had previously contracted with General Electric in 1944, in response for Germany's V-2 project.

On April 16 1946, the first V-2 rocket launched was from White Sands. From that point until September 19th 1952 the V-2s launched 67 to the New Mexico skies. Instead of explosive payloads these missiles were equipped with cameras, Geiger counters, as well as other instruments for science inside their noses. Mice, as well as Rhesus monkeys, too, were taken high to assess health risks associated with high speed travel at unimaginable high altitudes. On the 30th of July 1946, a rocket called V-2 attained an unimaginable height that was 100 miles. In the same year, another V-2 was an initial launcher that could detect the Ozone layer.

A post-World War II budget-conscious Congress was hesitant to allocate significant funds for research and development in rockets as well as the German scientists received a base salary of $144.00 monthly. Werner von Braun, himself was paid $9,500.00 annually, which included the benefit of $6.00 per day allowance while on the road. In pursuit of their dream of exploring space The majority of rocketeers refused to accept higher-paying private sector jobs and decided to remain civil service workers.

In 1946 in 1946, Major Colonel William E. Winterstern, caretaker of the German astronauts, asked an oblique query of von Braun: "If we could provide you with all the money you'd like and you had the time, how long would it have to travel into the Moon and return?" The rocket scientist was, at that time, unnoticed and sought some time to think about Winterstern's vast inquiry. After a few weeks, von Braun offered his reply: "Give us three billion dollars and ten years and then we'll take you back to Moon and return."

When as the United States established a fledgling missile program in the 1930s, it was the Soviet Union was busy developing and testing its own missiles. Similar to Wernher von Braun the Russian Sergei Korolev was a visionary who established the Group for Investigation of Reactive Motion in the 1930s. Korolev was appointed Colonel of the Red Army, traveled to Germany soon after the close of World War II and supervised the conscription of over 150 rocket researchers and technicians. In contrast to those in the United States, which allowed its German immigrants to play an active part in advancement and design, Soviets simply took lessons from their soldiers, and then sent them to their homelands. With only a handful of V-2s Soviet researchers rely on German concepts and Russian knowledge to design new missiles. Korolev would later be acknowledged as the chief designer of the Soviet space and missile programs.

There were few allies within its borders as well, it was the Soviet Union did not have accessible airfields from which they could carry out nuclear bomber strikes at United States. United States. To deter the powerful American nuclear bomber army, the Soviets were able to build nuclear missiles as a means of deterring.

When the Korean War in June of 1950, American defense spending, which had been severely cut in the aftermath of World War II, dramatically increased. While the war raged across the Asian peninsula and the peninsula was ravaged, the Army programme for missiles was further bolstered with money.

In the late 1950s, von Braun relocated to the Redstone Arsenal in Huntsville, Alabama, where he was appointed Director of the newly established Army Ordnance Rocket Center. Alongside the 115 other colleagues of his as well as their families as well as civil General Electric employees, and Army specialists in the fields of math, science as well as engineering. von Braun set to develop Redstone, Jupiter, and Pershing intercontinental ballistic missiles (ICBMS).

The arrival of Germans to Huntsville transformed the small North Alabama town into a center of study and research. The locals at first were unsure of what to think from their new neighbours and joked about calling their rapidly growing population in the form of "Hunnsville." Then, in March 1955, the German researchers, technicians, as well as their family members were admitted into the United States as U.S. citizens during a massive ceremony in Huntsville.

It was the Soviet Union, which had already set up the first launch location known as The State Central Test Range, was busy in the development of rocket-powered weaponry. The project was led by Sergei Korolev, in 1953 the Soviets revealed the R-7 rocket. With 20 kerosene-fueled and liquid oxygen burning engines, the robust R-7 could produce 1.1 million pounds thrust.

In 1955 the president Dwight D. Eisenhower, worried about the increasing Arms Race, proposed an Open Skies policy to the Soviet Union, whereby the two nations would utilize surveillance aircraft to observe the military development of each other. Soviet President Nikita Khrushchev immediately rejected Eisenhower's idea, believing that it was a lie that the U.S. was seeking a easy way to spy on its adversaries. In the meantime, Khrushchev was convinced that missile technology could allow for the Soviet Union to compete against the United States in the nuclear arms race. In 1956 the Soviet Premier known as a master bluffer declared that his nation was within the range of having "a guided missile that had an hydrogen warhead which could be launched anyplace around the world." After establishing an intimidating base in the world, this Cold War would dominate East/West relations for the next half century.

The 1950s were the time when the German-born rocket scientists of America developed V-2 technology and were making the Redstone rocket, the first ballistic medium-range missile, as well as this was also the car that would later send the first spacecrafts. Missiles that carried nuclear warheads but, were merely as a way to end in the case of Wernher Von Braun. A master of public relations, von Braun correctly sensed that the best method to promote his vision of space exploration was to appeal to the general populace. He written The Mars Project, a novel that told the tale of a trip into The Red Planet, stimulating the fascination of America's space-lovers. Between 1952 and 1954 Collier's journal featured an entire series of space exploration. Von Braun authored the first piece, titled Man Can Take on Space. In a later issue of the magazine that is widely read, von Braun predicted a human-manned mission to Mars will be possible in about 25 years. "There aren't any problems with which we don't have the answers or the means to find these answers right now."

In the last decade, the von Braun's lifetime wish would come true. It was expected that the United States and the Soviet Union will compete head-to-head in the multi-billion dollar race to discover space.

# Chapter 4: Beep-Beep

On the 4th of October, 1957, the entire world was abruptly and unintentionally exposed to Space Race. That day in the autumn of 1957, Americans were preoccupied with other pursuits. There was a World Series. New York Yankees and Milwaukee Brewers were tied, just one game apart in the World Series, while CBS viewers were anticipating the premiere of season one in Leave it to Beaver. At the point that anyone living in America United States was aware that the Soviet Union had launched a satellite, it was twice in orbit across North America.

Sputnik Zemlyi, "traveling companion of the world" was launched without much celebration, but the world's first satellite would drastically alter the dynamic of the superpowers. It orbited at more than 25 times that speed it appeared like an illuminating light in the night sky, a visible representation of the Soviet Union's head. beginning the exploration of space.

Sputnik (its initial name changed to Sputnik, but was eventually removed) was fitted with radios with a distinctive, beep tone could be heard by short-wave radio listeners across America. In an circular 141.7 588 mile orbit Sputnik's Soviet satellite was circling Earth at least once every 12 seconds for 96 minutes. Sputnik was on the moon, taunting rest of the world until January 1958 when the satellite ignited, and re-entered the Earth's atmosphere.

The president Dwight D. Eisenhower attempted to downplay his significance for the Sputnik launch by saying that the satellite was "one tiny ball floating in the air. It does not cause me to worry in the least." Eisenhower's view however was not the majority view and was evident by the remarks of strong Senate Majority Leader Lyndon B. Johnson: "The most important thing about the satellite is that it means that we cannot continue to consider that the Soviet Union to be a nation that was years behind us in research and industrial capacity." The charismatic Texan who was well-known to Washington in the know by the title of"Master of the Senate and the Master of the Senate, issued a loud warning to his fellow citizens that the Soviets were soon to "be dropping bombs at US from outer space just like kids who throw rocks over freeway crossings." A number of other powerful officials echoed Johnson's warning. The Washington Post likened the Sputnik launch to the Japanese attack on Pearl Harbor on December 7 in 1941. It was a moment at a time when there was a time when the United States had been caught completely unaware of the dangers.

Alarmists, who were clearly with the majority of them, said that the Russians were able to take an unbeatable advantage on the Space Race. Many military and political officials, who were all but exuberant, believed that about the Soviet Union would soon be launch missiles that were nuclear from the space. The British newspaper Manchester Guardian offered a grave but misguided warn: "Russians can now build ballistic missiles that can hit any desired target, anyplace around the globe." A confident and aggressive Nikita Khrushchev ignited the frenzied fires in the midst of in the Iron Curtain, boasting that the Soviet Union could launch nuclear missiles whenever and wherever it desired.

The deep psychological and practical consequences of Sputnik led to America's entrance to space. Space Race. Many politicians, scared of being seen in a negative light on Communism in the red-baiting Cold War era, exploited the anxieties of their constituents. On the 25th of November of the year 1957 Lyndon Johnson initiated congressional hearings to decide how best to boost the nation's fledgling space program. Within a short time, Johnson, who had his sights set on a presidential election in the year 1960, was named chairman of the Special Committee on Space and Astronauts. Other presidential hopefuls, including Kennedy, the Democratic junior senator of Massachusetts, John F. Kennedy and John F. Kennedy, amplified Johnson's request.

The program was led by engineer and scientist, Sergei Korolev, the Soviet Union was well on its way to creating an impressive space program. In 1955, the construction began at Baikonur Cosmodrome. Baikonur Cosmodrome in the Soviet-controlled Central Asian Republic of Kazakhstan. Guarded by a strong military force, Baikonur was a top secret "closed city" where , in totalitarian manner research, development and the implementation in development and implementation of the Soviet space program were kept secret from the outside world. It was only the camera of American U-2 spy aircrafts were aware of the operations at Baikonur. With its formidable R-7 rocket, which had already proved the capability to launch an orbiting satellite it was clear that it was clear that the Soviet Union was preparing to quit it's Cold War rival in the beginning blocks.

President Eisenhower always calm in any crisis, whether real or imagined and wars, was conscious the fact that Soviet missile technology had been massively overrated. He also knew that the missile gap was fiction based on political motives. The detailed photographs of U-2 flight over the Soviet Union had provided the President with proof that Khrushchev's dire warnings regarding the superiority of Russia's nuclear missiles were mostly in the form of boastful speech. In a bid to hide the secrets that was this U-2 project, Eisenhower played his top secret cards close to his vest and chose to ignore the insinuating assertions of the proponents of a missile gap.

In less than a month of its first triumph that year, it was just a month since the Soviet Union launched Sputnik 2 which is a satellite weighing 1,120 pounds (the dimensions of a tiny car) that carried a dog passenger into the orbit. Laika (Russian word for "barkerY) is the first dog to orbit in space can be seen barking on the satellite's radio transmitter. This is proof that living things can survive at least for a brief period in a space with no gravity. Despite growing concerns regarding the twin Soviet achievements and the American press slammed this satellite with the names Muttnik or Poochnik. Laika passed away after four days in orbit due to oxygen deprivation and overheating or from poison injections and the latter, which was what the Soviets considered to be more humane because the satellite couldn't be saved. Laika's demise enraged American pet owners, they found another reason to dislike those Godless Communists.

In the following three years over the next three years, over the next three years, Soviet Union would launch more Sputnik satellites. However, none would have the same awe-inspiring impact as the initial one.

The first time that America tried to participate in to Space Race ended in failure. While avoiding the expertise of German-born rocket scientists working at Redstone Arsenal in the army's Redstone Arsenal, the Eisenhower Administration chose to use navy's Vanguard rockets to launch the nation's first satellite orbiting. The anti-German bias persists in the administration's leadership and, as a result, they chose to use the "made by America" Navy rockets were offered the chance to become history. Werner von Braun shared his regrets: "This is not a design contest. It's a race to bring a satellite into orbit. We are ahead of the game on this."

On the 6th of December, 1957, a large group of photographers and reporters assembled in Cape Canaveral to witness the Vanguard rocket launch attempt, which was to launch an 3.2-pound satellite to orbit. The rocket was able to lift 4 feet from the ground before collapsing into an erupting ball of flame. The grapefruit-sized satellite managed to emerge from the fiery launch pad without injury and even prompted New York Journal American columnist Dorothy Kilgallen to say: "Why doesn't somebody go out and take it down?"

The media portrayed the failed launch in the form of Flopnik, Kaputnik, and Stayputnik. Von Braun found it increasingly difficult to control his anger: "...We could have had this happen using the help of our Redstone (missile) 2 years back... Vanguard will not be able to make it...We have the technology in the shelf...For God's sake, let our backs and let us try something."

As of the end of November in 1957 just one month prior to the Vanguard scandal the government had gave a green light to German rocket scientists from Redstone Arsenal, allowing them to join this satellite launch plan. A little over a month later, on January 31st, 1958, Explorer 1 was launched into space with an Juno rocket (a modified Redstone missile). Explorer 1 entered an elliptical 220x 15 miles orbit, speeding at 18,000 mph was able to circle the Earth each time for 114.8 minutes. This cylindrical American satellite, which measured 80 inches in length and weighing only 12 pounds, was ridiculed in the eyes of Soviet President Khrushchev as the "orbiting grapefruit." The American news media however they claimed the magnitude of the orbital landmark: "The 119 days between Sputnik 1 and Explorer were equally important for American history as U.S ....as anything else in a other similar period in the history of mankind."

Its Explorer 1 satellite proved to be an unqualified success it orbited its Earth more than 58,000 times prior to burning upon the return in March of 1970. The most notable accomplishments of the satellite were discovering the Van Allen Belt. With the help of sophisticated sensors, Explorer 1 verified that the radiation trap was circling the Earth at an altitude of more than 600 miles. This was a protection cover against potentially dangerous cosmic radiation.

The in-progress Vanguard satellite program experienced a second setback on the 25th of January 1958 in the event that the launch rocket's primary stage engine failed, just for 14 seconds prior to the it was ignited. On March 17th of the same year, the Vanguard satellite was successfully launched to orbit. The spacecraft which has measured Earth and Sun's gravitational field as well as solar winds and atmospheric conditions, is still in orbit to this day. The final successful Vanguard launch was in the month of March 1959, but the ailing Navy program's rate of success was not spectacular. only three of the eleven satellites were able to make it into orbit.

On the 29th of July, 1958 the President Dwight D. Eisenhower signed into law the legislation that established the National Aeronautics and Space Administration (NASA). This National Aeronautics as well as Space Act of 1958 not just established NASA however, it also specified that "activities in space must be used for peaceful purposes to benefit humanity in general." Keith Glennan, the president of the Case Institute of Technology, was named NASA's first Administrator. He received a clear order of the fiscally prudent Eisenhower to make sure that there was "no excessive spending."

The space program of America's infancy was put under the control of civilians. Alongside Wernher von Braun, and the Redstone Arsenal colleagues,

NASA also bought its own Army Jet Propulsion Laboratory, and the Navy Vanguard program.

In addition to committing to America's space program President Eisenhower promulgated to enact the National Defense Act on September 2nd the 2nd of September, 1958. In the following four years, close to billion dollars were allocated to provide low-interest loans to students who needed the profession of math and foreign languages teachers. The federal matching fund was allocated for private and public schools to purchase tools and equipment for teaching these subjects. Around 60 million dollars were allocated to finance 5500 graduate fellowships in engineering, science as well as foreign language.

On January 1st, 1960 the NASA's George C. Marshall Space Flight Center, named in the honor the president's World War II mentor, was officially opened in Huntsville Alabama's Redstone Arsenal. Wernher Von Braun was named director of the NASA facility, who was responsible for overseeing 5500 civil service workers and 1,189 contractors on site.

NASA has also created it's Goddard Space Flight Center in Greenbelt, Maryland. The new facility will assume the flight control and data retrieval responsibility of NASA's satellites that orbit.

In order to fully develop its research program in space, NASA needed an area for rocket launches. In 1946 NASA and the Pentagon determined a 15,000-acre tract of land in the Florida's Atlantic Coast as the ideal site for a launch location. This barrier isle, referred to by the name of Cape Canaveral, separated from the mainland by the Banana River, was isolated and remote, making it unlikely to expose civilians to the dangers that accidental explosives could cause. After NASA became a reality The Long Range Proving Ground at Cape Canaveral not only served as a missile testing facility but also as a launch location. The first space missions that were manned took off from a part that was Cape Canaveral known as ICBM row.

The impact of this on the socioeconomics of this remote part in the south Atlantic Coast was dramatic. Between the years 1950-68

Brevard County's population Brevard County, Florida increased from 23,000 to 239,000.

Cape Canaveral was ideally located to launch space launches from the east across the Atlantic Ocean, allowing rockets to follow the Earth's natural orbit. It is located close to the equator, at the point where Earth's speed of rotation is the highest (1,038 miles/hour) The East coast of Florida offered rockets that were launched with an advantage in getting the orbital speed 17500 miles/hour.

Even before the creation of NASA in the early years, the president Eisenhower didn't see space exploration as a major goal of his administration. He was puzzled with the "panic" created due to that Sputnik launch. A fiscally prudent man, Eisenhower enjoyed the fact that in his two terms of presidency America was able to avoid war and had a robust defense programwithout excessive spending. He was also determined to implement the same discipline for the Space program.

Although he was able to approve the first space program that was manned which was later renamed Project Mercury, Eisenhower refused to back the lunar landing programme (Apollo). Beware of the powerful and consuming military industrial complex, Eisenhower was hesitant to "hock his precious jewels" in order to make a human-powered lunar landing. NASA administrator Keith Glennan privately agreed with the Presidentthat "If we do not succeed in placing one person on the Moon in the next the 20th anniversary, nothing to lose."

In a bid to end the Republican Party's eight year control of the White House, the two contenders in the race for 1960's Democratic Presidential nomination John F. Kennedy and Lyndon B. Johnson, took up the issue of space and missile exploration to sway voters against the so-called sluggishness of the Eisenhower period. After Kennedy won the nomination and picked Johnson to be his running partner The Democratic ticket proclaimed the absence of a missile gap as proof that America's security was compromised through Soviet technological advancements. President Eisenhower's Science Advisor, James R. Killian, discussed this politically-motivated deception in his memoirs: "The drama of space stirred visions on the part of more than one politician that they might ride rockets to higher political ground."

In the year that John F. Kennedy was elected President in November 1960, NASA's space exploration program got an opportunity to breathe new life. In February of 1961, one month following his inauguration, the President was sworn in James Webb as NASA's new Administrator. A lawyer by trade and an experienced Oil company chief executive Webb was also Budget Director as well as under Secretary of State during the Truman Administration. With impressive organizational and business abilities the charismatic and strategically adept Webb successfully convinced Congress to allocate a substantial amount of money to NASA. In 1960, just a year prior to the appointment of Webb as NASA's Administrator, the agency's annual budget was set at 500 million dollars. By the year 1965, the amount was soaring by 5.2 million dollars. From 1960 until 1965 the number of workers at the space agency grew by 36,000 to 10,000.

The Soviets were able to remain at least a level ahead of United States during the early decades in space research. In 1959, on its own they Soviet Union sent three unmanned space probes to the Moon. In the month of January, Luna 1 became the first spacecraft that passed close to the Moon. It missed its destination by 3,700 miles Luna 1 entered into orbit around the Sun and took the first images of detail on the moon's surface. Luna 2, launched on the 12th of September, was intentionally crash-landed onto the lunar surface, making the first landing of spacecrafts in the Moon. As a result of a combination of science and propaganda, after impact, Luna 2 scattered metal pendants over moon's surface marking its space.

In October of 1959 Luna 3 was put into an elliptical high-altitude figure-eight Earth orbit. This drove the probe towards its Moon before returning. Luna 3's cameras captured its first pictures of moon's dark side. Soviets proudly christened two lunar marias with the names of"the" Sea of Moscow and the Sea of Dreams.

In a daring attempt to emulate Soviet Union's success, NASA initiated Project Mercury in the month of October, 1958. Abe Silverstein, Director of the Office of Space Flight Programs was the one who came up with the name Mercury as according to Greek mythology Mercury is the Olympic messenger was the child of Zeus and the grandson of Atlas.

The Mercury goal was to launch an unmanned spacecraft in orbit and return the astronaut safely home. Each mission would have one astronaut. The pilot would be taught how to maneuver the spacecraft and test how the body reacts to long-term experience with zero gravity. There are certain aerospace medicine experts were concerned that an extended period of weightlessness could be dangerous and possibly fatal for humans.

Specific criteria were set up for the selection of America's first astronauts. candidates needed to be less than 40 years old or less than 5' 11" tall (to easily fit in the cramped space capsule) in weight, not more than 180 pounds. and be active army pilots (with more than 1500 hours of jet-flight experience) and hold the bachelor's diploma or "the equivalent."

It was the Eisenhower Administration established the selection guidelines-to deter untrained and even untrained individuals from applying to be astronauts. Along with their unquestionable courage and skill, military pilots were also accustomed to obeying orders, something that appealed to Eisenhower who was the most famous World War II General. Many famous test pilots who gained fame for breaking new jet air speed records after World War II period were not included in an astronaut's training programme. Chuck Yeager, the first pilot to break through the audio barrier (Mach 1) was disqualified due to the fact that the fact that he did not attend college. Famous test pilots Scott Crossfield and Joe Walker were disqualified due to due to their status as civilians. They seemed not to be dissatisfied in their belief that space flight inside the "capsule" is not really a true test of a pilot's abilities and joked that astronauts could be nothing less than "Spam in a bottle."

Of the 554 active test pilots from the military Only 110 had met the criteria for astronaut selection. Based on evaluations from their commanders and flight instructors and flight instructors, the list was reduced to only 69. After being interrogated with NASA officials and being told about the requirements for NASA astronauts' training pilots decided not to take part. Out of the 32 pilots, 14 dropped out of the selection process after an intense series of physical, mental physical, and psychological tests. This left a list of 18 applicants.

Concerns about the risks to health from weightlessness and the effect of the powerful transverse forces during the lift-off phase and the re-entry phase of space flights resulted in NASA doctors to submit astronauts to a plethora of medical tests that, at times, did not have a rational explanation. The tests were performed by the Lovelace Clinic in Albuquerque, New Mexico, a medical facility that specializes in aerospace medical tests. John Glenn, who was one of the 18 finalist, recounted the experience in the Lovelace Clinic: "They drew blood, collected stool and urine samples as well as scraped our throats.

They took measurements of what was in our stomachs, provided us barium enemas and we were submerged in water tanks to count the body's total volume. They sputtered lights into our ears, eyes nostrils, noses and all over. They took measurements of our heart blood pressure, pulse rates as well as brain waves and the muscular responses to electrical current. The inspection of the lower part of the bowel was by far the most painful procedure I've encountered; an sigmoidal probe, paired with the device we were examining was nicknamed'steel. The tubes and wires were dangling from us like jellyfish's tentacles. We were not told what the strange tests were intended for."

The astronauts who were selected were given a variety of psychological tests. These included Rorschach ink blots as well as The Minnesota Multiphasic Personality Inventory, the Thematic Apperception Test, and the Guilford Zimmerman Spatial Visualization Test. Finding out if someone was suffering from psychopathy was of primary important to NASA officials who were cautious about being concerned about an NASA astronaut "cracking up" during space flight.

After spending eight days at the Lovelace Clinic, the prospective astronauts were taken for their Aeromedical Laboratories at Wright-Patterson Air Force Base in Dayton, Ohio, where their physical fitness was assessed using stationary bicycles, treadmills and by repeatedly walking between 20-inch tall boxes. While there astronauts were subjected to a second series of uncomfortable tests for their physiological health. Cold water was injected in the ears of test pilots to gauge movement of the eyes (nystagmus) and the feet of their test subjects were immersed buckets of ice to measure changes in blood pressure as well as pulse. To determine lung capacity, the test pilots were required to blow air into tubes and keep a column Mercury elevated to as long as they could. They were also put in heating chambers, and baked at temperatures up to 130 degrees F. To gauge the pilots' reactions to 65,000 feet altitudes They were also required to be confined in specially designed chambers and wear only pressurized suits that were partially compressed. Tilt tables were employed to gauge the degree of vertigo felt by every man, and high-speed centrifuges exposed them forces as high as G14 (G-1 equals the gravity pull on the Earth For every G or acceleration force a person is subjected a number of times of their body weight. A 175-pounder at G-3 will experience five times the force of 525 pounds). The pilots were then locked in a dark isolated chamber that was soundproof for three consecutive hours without being told beforehand, of the length of their deprivation of sensory stimulation. To test their ability in managing emergency situations, each pilot was tested on the "idiot" box, a device that has multiple simultaneously activated buzzers as well as flashing lights. Each candidate was tested to determine how quickly he could shut off the confusing array of visual and auditory alarms.

Seven pilots were selected to pilot Project Mercury-- Alan B. Shepard, Jr., Virgil I. "Gus" Grissom, Gordon Cooper, John Glenn, Scott Carpenter, Wally Schirra as well as Donald "Deke" Slayton. A wide range of Armed forces was represented-The list included: Glenn was an Marine; Shepard,

Carpenter as well as Schirra had been Navy pilots. Grissom, Slayton, and Cooper were Air Force men.

The Mercury 7 was recently christened. Mercury 7 made their public debut on April 9th 1959. The Mercury 7 sat behind a table in the Dolly Madison House in Washington D.C.'s Lafayette Square, within proximity to the White House, the astronauts dressed with civilian coats and ties were able to answer inquiries from reporters. Life magazine writer, John Dille described the young-looking astronauts "part pilot and engineer, part scientist as well as guinea-pigs and a heroic." The phrase "star voyagers" was soon a synonym for Mercury 7. Mercury 7.

The astronauts started their new jobs at the GS-12 civilian salary of $8,300.00 per year. However, their the new fame they gained brought some luck. For the benefit of Mercury 7, NASA Public Affairs Officer Walt Bonney, contacted Leo DeOrsey, a seasoned tax lawyer and personal representative of several film stars. DeOrsey was willing that he would represent astronauts free of charge for a fee, and then reached a three-year $500,000.00 agreement in partnership with Life magazine, whereby the popular magazine had exclusive right to biographies of astronauts as well as their wives. Each astronaut was compensated $23,809.52 in three monthly installments, almost the equivalent of doubling their annual salary. Life also provided men with life insurance who could otherwise be unattainable due to the risky nature of their jobs. Because of this agreement Life released more than 70 pieces of writing on The Mercury 7 as well as their families.

To fulfill their role as media sensations The former test pilots were instructed on the proper manner of dressing and posture, as well as speech-making. Michael Collins, who would be chosen in the next astronaut class, compared the school's curriculum to a charm class.

In every place the brave, clean-cut astronauts went were welcomed as America's most brilliant and best. Wernher Von Braun who was overseeing designing the rockets that carried astronauts to space, was impressed with their performance in the Mercury 7: "They are the most amazing group of individuals you've encountered. There are no daredevils in an incredibly long shot However, they are seriously sober committed and well-balanced individuals. ..."

In Space Race Space Race, image often overcame reality. The astronauts were separated from wives and their children in the arduous exercises in Cape Canaveral, and lured by the temptations offered by close by Cocoa Beach, a number of the Mercury astronauts were incongruous with their straight-laced images in the media. Their drinking habits, feminizing and high-speed races in their sports cars were removed from the contemporary lexicon of exploration in space. John Glenn, whose squeaky clean appearance secluded him from the scandals of Florida and elsewhere, confronted fellow astronauts over the rumors of their nefarious behavior. Being of the opinion that the line between duty and fun was clearly defined, Allen Shepard told Glenn with no ambiguity to cut off their private lives. fellow travelers.

As in the United States was initiating Project Mercury while Project Mercury was in the making, and Soviet Union was busy establishing its own space-based manned program. In the month of March, 1960, cosmonaut training started in the Star Town Facility, located only a few kilometers from Moscow. The candidates for cosmonaut training were chosen from the roster of active military pilots. To preserve the purity of their ethnicity, all the potential cosmonauts were required to be pureblooded Russians. In a strict manner, every Cosmonaut was required to show complete loyalty to his country and never doubt the decisions of his superiors even in flights.

From a pool that began with 20 pilots, 12 cosmonauts were picked and were dubbed Star Town 12. Star Town 12. The first class comprised Yuri Gagarin, Victor Gorbatko, German Titov, Georgi Situviv, Andrian Nikolayev, Yeugeni Khrunov Pavel Popovich, Boris Volyanov, Valeri Bykovksy, Aleksei Leonov, Vladimir Kumarov, and Pavel Balyayev.

The first human-powered Soviet spacecraft was known as Volstok (Russian which means "upward moving"). The spacecraft was designed by engineer Oleg G. Ivanovsky The spherical capsule was connected to a conical-shaped module that contained communications and telemetry equipment, nitrogen and oxygen tanks, antennae and retrorockets. As the spacecraft returned home it was expected that the equipment module be removed before the spacecraft returned to the atmosphere of Earth, and the capsule would then parachute for an earth landing. To ensure safety, on the first space missions the cosmonauts were instructed to fall from the capsule before the capsule hit the ground. In a show of bravado, Soviet authorities disingenuously reported that every cosmonaut was on the spacecraft until it came to rest.

In the beginning, America's first astronauts were subjected to one of the potential disasters related to space flight. The 18th of May, 1959 The Mercury 7 witnessed a test of the Atlas rocket that was to send them into space. After launch, the huge rock to a halt within the Cape Canaveral skies. John Glenn likened the blast to an "hydrogen bomb that went through the heads of us." In a short period of stillness, Alan Shepard eased the tension by displaying his morbid comedy

"Well it's good that they were able to get that out of out of the way."

Second non-manned Mercury mission, which was launched on the 21st of November, 1960 failed as well as the Redstone rocket lifted just several inches from the launch pad, it crashed into the supporting tower. The rocket was teetering on the edge of danger and the capsule's parachutes flew out like a party favour--an embarrassing display for what was popularly known in the media as"the "four-inch flight."

On December 19th the similar year NASA had the opportunity to send an spacecraft that was not manned MR-1 rocket into orbit. After reaching an altitude of 131 miles the spacecraft returned safely to Earth after splashing into the Atlantic Ocean.

To prepare for the spacecraft's manned flight, NASA experimented with animals. The first participants included pigs. They were shackled inside Mercury capsules, and then suspended from high altitudes in order to test the spacecraft's resistance to impact. The pigs emerged from the experience with just minor injuries that humourously validated the assertions of many veterans of test pilots. Astronauts were only "Spam in the Can."

On the 29th of May 1959 NASA released the Army Jupiter missile carrying Able, the rhesus monkey Baker who was a South American squirrel monkey, into space. Both monkeys were fitted with electrodes to gauge their physiological responses to G-forces and weightlessness during the 300-mile flight. After the splash the passenger compartment of the rocket was found in within the Atlantic Ocean, where both passengers were discovered tucked away in a safe and secure location.

The next space explorers to be discovered were the chimpanzees. They were chosen due to their reactions times being almost identical to humans. An entire group of 40 chimps in the New Mexico's Holloman Aerospace Medical Center were prepared for space flight. On the 21st of January, 1961, six Astrochimps, as well as 20 handlers and medical experts traveled between New Mexico to Cape Canaveral to prepare for the first launch of test flights. The now irritable Mercury astronauts had doubts about the need to conduct more tests and a disgruntled Alan Shepard expressed hope that the next test launch would result in the creation of a "chimp grill."

On the 31st of January in 1961, the ape of 61, affectionately referred to as Ham, an abbreviation that stands for Holloman Aerospace Medical Center, was launched into space. In the cockpit of an inflatable pressure chamber made of plastic that was that was the size of the trunk (designed to replicate the environment of the suit of space), Ham endured the 16-minute, 39-second space flight suffering only minor injuries, a bruised nose which occurred in the course of lift-off or splash down.

Ham's mission was not without errors. The spacecraft's retrorockets got removed too early, thereby increasing the speed of re-entry by 1,400 miles/hour, the craft crashed down 130 miles over the intended zone. After impact there were two holes punched into the capsule, which caused it to take on an estimated 800 tonnes of ocean water. It took more than two hours to Navy helicopters to find the capsule that was listed; by the time they arrived, Ham was in a fury, screaming and biting at the rescuers. At the post-flight press conference cameras flashed also irritated the chimp which flinched his fangs in front of the world.

As monkeys flew through space, Mercury 7 proceeded with training exercises. The astronauts were taken on flight aboard F-100 jets, performing Mach 1.4 dives and C-130 transport planes that flew parabolas. Both exercises allowed them to experience the sensation of being weightless for a while.

Utilizing flight simulators, astronauts were able to familiarize themselves with the brand new spacecraft. In the beginning of January 1959 McDonnell Aircraft Corporation had been given the contract to construct twenty Mercury capsules. More than 4,000 companies ultimately provided parts or supplies to build the spacecraft.

Just six feet, ten inches long and six-feet-two-inches-wide (at its widest) The 4,300-pound spacecraft's interior was congested. The flight technicians and engineers however, were not worried about astronauts' comfort. They saw the pilots as unnecessary components to space flight. The spacecraft's propulsion and altitude guidance as well as re-entry mechanisms were conceived to be operated by ground-based techs. The astronauts were furious at the diminution of the role of the pilot, and Deke Slayton remarked: "Mercury was designed to operate without pilots." At some point there was a thought of administering drugs to astronauts before launch, rendering them invulnerable to the effects of space sickness and G-force pain. It also prevented the astronauts from pressing buttons or turning off the switch in the cockpit.

"All we require to flimse everything up would be a competent space pilot with hands eager to take control," a Bell Lab engineer grumbled.

In the final analysis, NASA needed heroes as the spacecraft did even if it was for nothing other than propaganda. In dissatisfaction with their position as passive capsule occupants they Mercury 7 successfully lobbied to alter the spacecraft, which included installing a manual backup navigator, cockpit windows as well as an escape hatch equipped with bolts that were explosive. This feature was thought to be necessary, given that astronauts were not willing to depend on anyone else to rescue them from the spacecraft, in the situation in an emergency.

Before an unmanned Mercury spacecraft was ever launched and landed, it was the Soviet Union scored another first. The 12th of April, 1961 Cosmonaut Yuri A. Gagarin was the first to be the first to orbit the Earth. An ex-fighter pilot at the age of 27, Gagarin accomplished a single orbit lasting 148 minutes and 1 hour onboard the Vostok 1 spacecraft. While on space Gagarin Soviet astronaut ate, drank and wrote notes on a pad, which proved that the metabolic, digestive and neurological functions weren't significantly affected by the lack of weight.

When he returned on Earth, Gagarin earned effusive praise from Nikita Khrushchev "You have created you immortal." In addition to fueling the fire of propaganda, Gagarin boasted: "Let the Capitalist nations get ahead of us."

It was reported that the Soviet journal along with the Communist mouthpiece Pravda claimed that Gagarin's space mission was an "great occasion within the history of mankind." In the same at the same time, The Washington Post echoed the angst of a lot of Americans: "The fact of the Soviet space flight has to be acknowledged for what it really is, and it's a psychological victory of the highest order to this nation, the Soviet Union." In over the course of 26 months the Soviets succeeded in launching five more space missions that were manned which convinced many that America was stuck in second position within Space Race. Space Race.

Following the failure of the retrorocket in the chimp's flight Wernher von Braun pushed for another unmanned test flight much to the displeasure and dismay of Mercury astronauts. The test flight was successful, as the spacecraft followed the correct path and landing 307 miles to the south of distance in the Atlantic Ocean. Be wary, NASA officials had originally thought of sending more chimps into space but when that the Soviets sent Yuri Gagarin into orbit, the United States was pressured into creating its own spacecraft with manned pilots.

On May 5th, 1961 Alan B. Shepard, Jr. was the very first American to go into space. The morning before this historic flight, Shepard was eating breakfast consisting of filet Mignon, scrambled eggs and orange juice before wearing cotton underwear and a spacesuit. The suit, which was manufactured at B. F. Goodrich in Akron, Ohio, was composed of aluminized plastic and nylon. It was it was a modified model of the Navy Mark IV pressurized suit. In the space suit, Shepard's temperature was maintained and his body's smell was eliminated through the activated charcoal filters. Oxygen that was vital for life entered the clothing at the thorax, and was released via the helmet. For his official launch attire, Shepard wore custom-designed gloves with boots, gloves, and a helmet. The whole 22-pound suit cost $5,000.00.

Over the launch pad and assisted by backup Pilot John Glenn, Shepard squeezed into the space capsule of 4,300 pounds named Freedom 7; each Mercury spacecraft would have the same number in honor of America's initial seven astronauts. In the cloudy skies the half-Moon was obliterated Cape Canaveral, as Shepard stood on the Redstone rocket which was an improved version of the famous V-2 that was capable of producing 37,000 pounds thrust. Two movie cameras were positioned inside the capsule, one to watch the instrument panel, and the other was used to record the emotional and physiological responses of Shepard during the space mission.

The other astronaut Deke Slayton, who was stationed in the control centre was named the cap communicator (Cap Com)--the person who maintained the radio contact to Shepard during the course of his flight. Before the launch, Slayton was joined at launch control by John Glenn and Gus Grissom. In a nearby building, Gordon Cooper monitored weather conditions and was on standby in coordination with rescue teams in the case in the event of an emergency. Wally Schirra and Scott Carpenter were waiting in the close by Patrick Air Force Base; they were secured in their cockpits of F-106 jets, positioned to pursue the spacecraft following its launch.

Another man who's name would be associated to space travel, was present on Cape Canaveral to witness Shepard's historic launch. The telecasting industry was in its infancy in terms of technology as was CBS News anchorman, Walter Cronkite was required to read the telecast from behind of the station wagon which was in sight from the pad for launch.

A 44 year old Cronkite was a fervent space lover. NASA made use of media attention to increase the size of its space program and Cronkite was one of the most sought-after spokespersons. The space agency was designated as an insider, the journalist was in the know about specific details regarding space missions and had personal connections with a number of astronauts.

Cronkite's rise to fame in the 1960's coincided with the growth that was his involvement in the American Space program. In the final years of his career Cronkite would be referred to for being "America's anchorman" as well as the "most respected person on the planet in America." Certain journalists, however were critical of Cronkite's support of space exploration, claiming that he was defying his journalistic independence to be an "cheerleader" to NASA.

Cronkite's enthusiasm was evident in CBS broadcasts and he made no attempt to apologize or excuse his regularly humorous and smug comments. Cronkite admitted his part in the celebration and praising of Mercury 7 astronauts: "We were well conscious that the image NASA tried to portray was not entirely truthful. However, at the same it was a recognition that our nation required heroes."

Prior to the take-off of Freedom 7, Alan Shepard was confronted with one of the most basic human need. Since the first Mercury flights were of such a brief duration, the spacecrafts did not have toilet facilities. The Freedom 7 launch was delayed multiple times, Shepard was forced to use his space suit to urinate; NASA medics shut down the electrical sensors to prevent a electrical short. While he waited for the final countdown, America's very first space explorer was contemplative: "I just kept looking at me and recollecting that everything inside the capsule was provided by the bidder with the lowest price."

The countdown for blast-off was frequently delayed due to clouds as well as an inverter that had been overheated that needed to be repaired. Squeezed into his cramped cockpit sofa, Shepard's annoyance was escalating: "Why don't you fix this small issue, and then light the candle?"

In the 9th minute of 9:34 a.m. when 45 million Americans were watching on television the massive Redstone engine finally came and the ground shaken. At the bottom of the rocket, strong steel flame deflectors were designed to channel their exhaust from engines. The water streams, which flowed at a rate of 35,000 gallons every minute, chilled the deflectors, generating massive steam clouds which partly obscured the pad.

As the rocket rocket rocketed upwards across the Atlantic Ocean, Shepard radioed launch control: "Roger. Lift off and the clock has been started."

Eighty-eight seconds after blasting off the rocket was able to surpass Mach 1 and eventually reached the speed that was 5,100 miles an hour. Through the periscope on the capsule, Shepard was able to discern Florida's west coast and the Gulf of Mexico, mammoth Lake Okeechobee, and the Bahamas.

The 15-minute flight of Shepard's followed an arc of 302 miles, reaching an peak of 116.5 miles. It was sunk by the Atlantic Ocean. He was enduring five minutes of gravity-less forces and weightlessness prior to landing at 260 miles to the south in Cape Canaveral.

In the course of his historic flight Shepard briefly turned off the autopilot, and instead utilized his manual stick for control. Hydrogen peroxide jets, emanating from nozzles located on the sides that of the capsule, let him to test every axis of flight, including the pitch, yaw as well as the roll and pitch of the spacecraft. Shepard discovered the short period of being weightless "pleasant and relaxing" partly dispelling the concerns of a lot of NASA medical experts.

In the midst of the fiery re-entry into Earth's atmosphere, the outer wall of Freedom 7 soared to 3,000 degrees Fahrenheit. Although the inside of the spacecraft reached temperatures of 100° (F) however, Shepard's insulated space suit was never higher than the temperature of 82 degrees (F). At 10,000 feet the spacecraft's main parachutes was released, and a few just a few minutes after, Shepard likened the impact of the splash as the impact of landing an aircraft onto the surface on an air carrier.

Alan Shepard's first space mission made him an immediate hero. He was a hero to millions. Mercury 7 astronauts as well as their spouses attended a reception at the White House, where President Kennedy gave Shepard with the Distinguished Service Medal during a ceremony held in the Rose Garden. Around a quarter million people filled in the city streets New York City during a parade with ticker tape to celebrate the nation's first star-voyager. It was the Freedom 7 capsule was sent to be displayed at the Paris Air Show. It was the first time in history, it appeared that the United States appeared to be moving forward within space. Space Race.

On the 10th of April 2011, three weeks prior to Alan Shepard's first flight in history The President Kennedy persuaded Congress to change the Space Act, which had been approved under the Eisenhower Administration. Kennedy asked that the Vice President rather than the

President, be Chairperson of the Space Council. Since his early days, he has been a passionate advocate for exploring space, Lyndon Johnson readily embraced his new role.

On the 20th of April, just 8 days following the day that Yuri Gagarin became the first man to orbit Earth, Kennedy charged his Vice-President to answer these question: "Do we have a chance of beating the Soviets by establishing an experiment in space or by taking a trip to the Moon or by using the launch of a rocket that would travel back to the Moon and return with an individual? Do you know of any other program with dramatic outcomes with the possibility of winning?" Emphasizing the words beat and win, Kennedy affirmed that the Space Race was as much technical as it was political.

After forming a special committee which was consciously populated with proponents in space research, Vice-President came up with a response to President's questions. On April 28, eight days after completing his task, Johnson presented Kennedy with an over-the-top, yet profound note: "Other nations, regardless of their respect for our idealistic principles tend to be a part of the nation they think is the leader in the world and will be the winners in the long run. Space-related achievements are increasingly being recognized as a significant indicator of global leadership...If we do not put in the effort now to make a difference to do so, the day will be upon us where the margin of sovereignty over space as well as mankind's minds due to space achievements will be so heavily towards one side, the Russian side that we'll not be capable of catching the pace, let alone take over the role of leader."

The 25th of May, just two weeks after Alan Shepard's first space flight, leveraging the national pride and overwhelming sense of achievement, Kennedy took his case directly to the American public. Speaking to a joint session Congress regarding "urgent national requirements," the President issued an uncompromising proposition: "I believe that this nation must be committed to accomplishing the goal prior to the time this decade of landing a human on the Moon and then returning him safely back on Earth." At the conclusion of his speech, Kennedy declared that his idea will be expensive:

"Let it be clear that this is a choice that the members of Congress have to take. Let it be known that I'm seeking Congress to make a firm agreement to a new strategy for action. It is a plan that will continue for a long time and will be very costly, with 531 million dollars for fiscal year 1962 as well as an estimate of 7 to 9 billion in the next five years. ..."

The proposal was met with opposition from a few Republicans who opposed it, such as senators Barry Goldwater and Representative Gerald Ford as well as Conservative Southern Democrats, like Senators Richard Russell and J. William Fulbright and J. William Fulbright, all of who opposed such massive spending, the President managed to an agreement in the eyes of both the American public as well as a large majority of legislators. After having experienced the lowest moment of his presidency a month prior, in the Bay of Pigs fiasco, Kennedy was able to capitalize on an issue that inspired his fellow Americans and re-energized his political fortunes. The traumatic effect of the idea of a Soviet Cosmonaut orbiting the Earth upon the American consciousness was huge and was reflected in the historian of space, Gerard J. Degroot's concise assessment: "Gagarin was Kennedy's Sputnik."

Kennedy's speech was inspirational However, many wondered whether it was attainable. With just one space mission under its belt, NASA had less than nine years to achieve JFK's vision. On 21 July 1961 President Kennedy signed into law the newly approved Extended Space Program Act, which was the beginning of Project Apollo.

In the early hours of 7:20 a.m. on July 21 in 1961 Virgil I. "Gus" Grissom became the second American to launch into space. After recalling Alan Shepard's bladder troubles in the initial Mercury flight, Grissom decided to put on an girdle made of women's underwear underneath his space suit, thinking it would help absorb urine in the event that nature calls. In the same vein, Grissom, like his colleagues in space, ate the diet with a low amount of residue for three days before their spacecraft launches, thus being able to avoid the urge to urinate during space travel.

Grissom's 15-minuteand 37-second suborbital flight on Liberty Bell 7 reached an peak of 118.2 miles and a top velocity of 5168 miles an hour. Grissom was able to experience 10 minutes of weightlessness, and did not suffer any adverse side effects. In response to astronauts' earlier requests, Liberty Bell 7 was equipped with an expanded cockpit window that provided Grissom with a clearer perspective during his space journey. Following the re-entry of the capsule, its main parachute was opened at 12,000 feet. seven minutes later, the spacecraft came down without incident, concluding what appeared to be an uninhibited mission.

While the spacecraft bobbed across the Atlantic Ocean, 302.8 miles away of Cape Canaveral, Grissom removed his helmet and pulled off his belt, then completing an after-flight checklist and contacting 2 nearby Navy Sikorsky helicopters to come to his aid. Without warning the capsule's escape hatch burst.

The activation of the escape hatch that was similar like the cockpits in aircrafts that Grissom flew and commanded the pilot to remove the lock pin before applying 5-6 tons of force to the plunger that exploded 70 bolts and blew the doors 25 feet of the craft. For the rest period of his existence, Grissom would insist that the plunger was not deliberately or by accident, press the plunger "I was concentrating on my own business and then I heard the sluggish noise."

Seawater was leaking into the capsule through the large hatch which forced Grissom to get out of the sinking vessel. Two rescue helicopters was able to grab this capsule but they could not lift the massive water-filled spacecraft. It was forced to end the rescue effort. Liberty Bell 7 quickly sank 17,000 feet down to below the surface of Atlantic Ocean, and would not be found for 37 years.

The helicopter crew was focused on salvaging their capsules, Grissom nearly drowned in the turbulent waters which were made more rough due to the downward draft of the choppers and rotating rotors. The spacesuit's buoyancy became heavy as the seawater flowed in through the neck's opening and the oxygen port which was unplugged by mistake. While he struggled to keep his ship afloat the spaceman became frustrated and angry at the rescuers' efforts to save their spacecraft. In the end some of the helicopter crew dropped a harness, and then retrieved the exhausted and soaked Grissom from the sea.

Grissom's explanation of the sudden detonation of the escape door did not go down well with numerous NASA officials. When reconstructions were made from the sinking spacecraft incident engineers could not open the hatch with no human intervention which led many to believe that Grissom had pressed the plunger for detonation by accident or with a motive to panic.

On the 6th of August in 1961 Soviet astronaut German Titov started a 24-hour space journey that orbited Earth 17 times. The cosmonaut tried out his Vostok 2 spacecraft's manual control of altitude and was the first to take photographs of Earth in space. Titov also indulged in a lavish dinner of liver pate, bread and peas, all as well as drinking Black currant juice. While not revealed by the sly Russians, Titov suffered severe motion sickness that partially decreased after he slept for five orbits. Similar to Yuri Gagarin before him, Titov was a hero. Titov was elevated by rank from major to captain.

With two suborbital missions under their belts, NASA was prepared to launch the spacecraft into orbit. The 13th of September, 1961 the MA-4 Atlas rocket launched a fictional person into space. The spacecraft orbited Earth twice before reaching the Atlantic Ocean target zone, three hours after lift-off.

Another chimpanzee, Enos took off into the space the 29th that year. Enos flight lasted three hours while he circled the Earth twice and experienced 181 minutes of in weightlessness. As with his counterpart in suborbital space, Ham, Enos was trained to carry out his duties in the cockpit on the basis of operant conditioning. If the chimp pulled appropriate levers, he was awarded with banana-flavored water and pellets. If he was unable to do the task properly painless electric shocks were injected into his soles. In the course of the orbital flight an error in the capsule caused the chimp to be strengthened by electric shocks regardless of the lever was pulled. When rescuers arrived at their capsules within the Pacific Ocean, a terrified and furious Enos had in pain ripped out his urinary catheter that was still inflated and ripped off his biomedical electrodes. It's not surprising that the furious animal tried to bite the Navy rescuers. In addition after the flight press conference, the chimp tore off his diaper, and then began kissing himself in front of TV cameras and reporters which earned him the undeserved moniker, Enos the Penis.

After 10 delays due to weather-related issues and equipment failures, John Glenn undertook America's first space mission in orbit on February 20, 1962. The more robust Atlas rocket was utilized as the initial launch vehicle to send the Mercury capsule into space. It was built by two California-based firms, General Dynamics and Rocketdyne The rocket was able to generate three times the thrust of (compared with the redstone's 76,000-pound thrust).

The job of putting an astronaut on orbit and returning him in safety was more risky than the two previous Mercury missions. If tragedy struck Kennedy had prepared a formal statement. Kennedy made a pre-planned formal declaration: "To Mrs. Glenn and all members from the Glenn family, I offer my sincerest condolences. It was an honor to have met John Glenn. This country and the whole world mourn to his family. Glenn family. Space scientists will remember his spirit of discovery for eternity."

at 9:47 a.m. After an interval of two days and 17 mins, Friendship 7 lifted off the Cape Canaveral launch pad; the historic event was watched to by more than 40 million viewers on television. NASA had set up communications stations throughout the world in order that Glenn could keep regular radio communication with Earth. From east to west the ground stations were situated in Cape Canaveral, Bermuda, the Canary Islands, Nigeria, Zanzibar and Zanzibar, a Navy vessel located in the Indian Ocean, Australia, Canton Island, Hawaii, the California coast as well as California coast, White Sands Proving Ground, Mexico, Corpus Christi, and Eglin Air Force Base (in the Florida panhandle). The operation of the stations worldwide was a huge undertaking comprising more than 19,000 people.

In the span of four hours and 56 second, moving at a speed 25730 feet per minute Glenn completed three full orbits. With this velocity, the astronaut observed numerous transitions from day to night and discovered that every day in space lasted less than 45 minutes. While in space Glenn took in portions of two meals in a row, proving that the metabolic and digestive processes worked in an unweightless space. On the first orbit the altitude control system of the spacecraft was not working properly, causing the spacecraft to drift off its course. By switching to manual control Glenn was able to fix the flight path that was off.

A shocking warning of the dangers that come with space travel occurred shortly before the end of Glenn's initial full orbit. Friendship 7's alarm system alerted that the spacecraft's ablative heating shield and the compressed landing pack had not been engaged to the lock position. If the shield didn't remain in its place, the spacecraft will begin to burn during the re-entry process. NASA flight controllers didn't completely inform Glenn about the severity of the situation, which is an unacceptable error to an skilled pilot. Instead, ground controls instructed Glenn not to remove the retrorockets that were used to place the spacecraft in the proper angle for re-entry into the atmosphere. They also hoped that the straps of steel that held the rocket pack would keep the heat shield into the right position. In the midst of the fiery re-entry procedure where friction caused by the extreme temperatures blocked radio signals between the spacecraft and ground controllers NASA officials held their fingers crossed hoping that the heat shield would stay in position. After several minutes of trepidation Glenn's voice could be heard on the radio, which confirmed an uninjured re-entry. The retrorockets and the straps supporting them had caught fire however, the ablative heat shield stood in place and sheltered the capsule in the event of re-entry. The furious Glenn supported by his fellow astronauts, demanded that NASA ensure that crew members were kept informed of any equipment malfunctions that might occur in the future mission.

John Glenn returned to Earth with a greater fan following that Alan Shepard. When Glenn was in Washington D.C., 250,000 people crowded Pennsylvania Avenue to watch him go by. After receiving a standing ovation in the White House, Glenn addressed the joint session of Congress. It was the Friendship 7 space capsule departed for a world tour, informing the world about America's milestone in space exploration.

Straight-laced, clean-cut veteran Marine was the ideal spokesperson of space exploration for the American NASA space programme. A charismatic, well-spoken, with great communication abilities, Glenn proved more valuable as a publicist than a pilot and President Kennedy was quick to remove him from the flight list. The historian, William E. Burrows, perhaps best summarized the situation: "John Glenn came out of the ashes of Friendship 7 to be the Lindbergh from his day."

Over the fifteen months following there were three more Mercury spacecrafts have been launched to space and the length of orbital flights gradually increased. Deke Slayton was the last on the flight list but a cardiac arrhythmia (atrial fibrillation) forced him to withdraw from the rotation. Slayton was deeply disappointed with the medical decision to disqualify him was then rewarded with the appointment to a new job in the office of the Astronaut Activities Coordinator that kept him engaged in future space missions.

On the 24th of May in 1962 Scott Carpenter was launched into orbit on Aurora 7. Carpenter was able to circle three times around the Earth three times, and performed experiments with liquids in an unweightless environment. A glitch in the automatic flight control system caused Carpenter to take control of the spacecraft. While in the cockpit, Carpenter burned fuel much more quickly than he anticipated and, at the moment of re-entry failed to trigger the retrorockets in the correct time, which resulted in a splash of 250 miles out of the zone of interest. In a state of panic, NASA officials as well as television viewers were waiting for over an hour until Navy rescuers found the spacecraft. Carpenter was found in a life-raft tied by Aurora 7, the latter of that was at risk of sinking.

At his post-flight press conference, Carpenter made the mistake of causing embarrassment to NASA by bringing attention to the long period of time that rescuers took to locate and retrieve the spacecraft: "I didn't know where I was and they didn't." NASA officials were already irritated with the fact that Carpenter was wasting fuel and had misjudged his re-entry and re-entry, which led Director of Launch Control Christopher Craft to grouse: "That little bitch won't fly for me ever again!" Craft's angry declaration proved to be accurate; Carpenter never flew again in space.

A few months later, Sigma 7 was launched into orbit. Wally Schirra executed 6 full orbits, which was twice more than Scott Carpenter, yet consumed less fuel than his predecessor. Schirra also landed precisely on time, and only 4.5 kilometers from the aircraft rescue carrier. NASA officials called it an "textbook flight."

On May 15th in 1963 Faith 7, piloted by Gordon Cooper, was launched into orbit. Cooper set a new record for space travel, 22 orbits over the course time of 34 hours, 22 mins, as well as traveling 546,167 miles. Cooper was also the very first spaceman to go to sleep while on space. Faith 7 was equipped with TV cameras, which broadcast some of the first real-time orbital images to those watching below. In the event that the spacecraft's auto-control system failed, Cooper had to use the control stick to hold the spacecraft steadywhile firing the retrorockets for return. Cooper's cool manner of conduct and his piloting abilities helped him avoid the catastrophe.

While Project Mercury wound down, NASA was already preparing for the next stage in space-based exploration. In the end, nine new astronauts were hired in 1962, and then 14 more in 1963.

While the attention of the world's population was focused to Project Mercury, NASA's highly efficient space exploration program without a pilot progressed. It was launched on August 7, 1959, Explorer 6 became the first spacecraft to take photographs of Earth from space. Pioneer 5, launched on March 11, 1960, sailed in orbit about the Sun and orbited located between Earth and Venus and became one of the spacecraft's first to study magnetic fields that exist between the two planets.

First launched on April 1, 1960, Tiros I, the first satellite to be launched in the world equipped with infrared technology for observation and television cameras. The in the same time, the world's first Navigation satellite Transit 1B, was launched into orbit and allowed American vessels at sea to determine their position with incredible accuracy. About four and a half months after that, the first satellite for communications that was not a test, Echo I, began orbiting the Earth. The satellite was described as a passive satellite for communications, Echo I functioned as an reflector, but was not a transmitter. The signals could only be transmitted to it, and then "bounced to return" towards Earth.

Through its successful non-manned spaceflight program as well as the advent of cutting-edge technological advancements, America was becoming more than competitive in the Space Race. In October of 1960 it was reported that it was reported that the U.S. had successfully launched 26 satellites into orbit. In addition, the success rate of NASA was dramatically increasing by the year 1958. All four launches failed. Nine of 14 launched successfully during 1960 and twelve out of 17 satellites made it to orbit in the year 1961.

Space intelligence-gathering took on a new meaning in the early 1960s, when CIA operatives "kidnapped" a Soviet Luna probe, while the spacecraft was being displayed at a trade fair in Mexico. The American agents kept the probe for a night, securing the vehicle, meticulously photographing the spacecraft, and then copying serial numbers of its major parts.

In the same time, American spy satellites were often taking photos of Soviet defense installations. The film was taken down by satellites and later retrieved through Air Force planes using hooking devices that snagged the photo packs that drifted downward with their parachutes.

As the decade progressed, America continued to refine its satellite technology. Telstar I, conceived in collaboration with Bell Labs and AT&T, was launched into orbit on the 10th of July in 1962. It was powered by solar cells. the satellite broadcast live television broadcasts to Europe and the

United States and Europe. Telstar I was the genesis of the more advanced communications satellites Telstar II, Relay, and Syncom. In July 1963 Syncom II was placed in geostationary orbit. It was so that signals could bounce between the two satellites from Earth which led to the term that has become a household phrase: "Live by satellite."

Geostationary orbits have increased satellite communication. Geostationary orbits are when satellites assume an orbit around the sun that is directly above the Equator and is able to follow the natural course of Earth's rotation. With an orbital duration that is equal to Earth's rotating time and a satellite that appears to be stationary at the top of the space (a fixed footprint). Nowadays the majority of communications and weather satellites are located in geostationary orbitsthat allow Earth-based antennas to remain in the same spot at the top of the sky.

Thanks to advances of rocket propulsion technology, navigation and navigation technologies the idea of exploring outer space was realized. The 27th August of 1962 NASA released Mariner 2, which became the first probe that flew straight to another world (Venus).

NASA did however have to contend with its initial moon exploration satellite. On the 23rd of April of 1962 Ranger 4 blasted into space, becoming the initial American spacecraft to touch the lunar surface. The initial and the second Ranger probes were stuck on Earth's orbits, after their engines in the upper stage failed and Ranger 3's upper stage engine sped for too long, which caused the probe to be missed by the Moon by about 22,000 miles. Unfortunately, Ranger 4 lost power following a crash on the lunar surface and was unable to send images as well as other important data back to Earth.

In September of 1962 the President John F. Kennedy delivered an unforgettable speech during a speech at Rice Stadium in Houston, Texas. Kennedy stated that America's aim was getting a human on the Moon prior to the date of the end of the decade "Some have asked what is the reason to go towards the Moon? You could also consider"Why climb the tallest mountain? Why do you sail on the ocean with the most expansive?"

Although many believed JFK's plans were a bit unrealistic The man who was who was responsible for the design of space launch rockets never wavered in his belief. Wernher of Braun who was at the helm for the George C. Marshall Spaceflight Center at Redstone Arsenal, was certain mankind would soon be on the Moon.

Von Braun was actively focused on the development of an Saturn V rocket--the launch vehicle that would take astronauts onto the Moon. Even before those outside the Massachusetts state Massachusetts knew about John F. Kennedy, Von Braun's plans for the long term were already written in the stone. In the beginning of May 1950 the Huntsville Times had informed its readers that DR. VON BRAUN SAYS ROCKET FLIGHTS POSSIBLE TO MOON.

For humans to reach the Moon in the future, two distinct spacecraft, but connected to each other, would need been developed: a vehicle that could journey to Moon and back, and another one that would be able to land upon the lunar surface. Astronauts must master docking and rendezvous maneuvers, and be able to control the spacecraft. With these lofty objectives in mind, Project Gemini was formally announced in the December of 1961.

In the beginning, it was referred to as Mercury Mark II, the Gemini spacecraft, built through McDonnell Aircraft, was considerably larger than the predecessor and comprised three distinct parts: a cockpit capsule for astronauts, an equipment module, which contained the power supply system and propellant tanks, communicationsequipment for instrumentation and communication, the drinking water tank; as well as finally, the engine compartment.

The significance associated with the most recent NASA project was evident. It is believed that in Greek mythology Castor or Pollux represented the Gemini twins. They were also one of the twelve Zodiac constellations which were ruled by Mercury. As with their namesakes Gemini spacecrafts could hold two-man crews, and follow in the tradition of Mercury.

To prepare for Gemini space flights and the future of space exploration NASA's

Manned Spacecraft Center relocated from Langley Field, Virginia to Houston, Texas, giving the birth of Mission Control. Although all space launches will be launched out of Cape Canaveral, Mission Control will serve as the primary center for all flights operations.

Texas is just happened to be the in the country of one NASA's most ardent supporters, Vice president Lyndon Johnson. Like most states, politics correlated with the money trail, but in a grander way with this Lone Star State. Humble Oil donated a large land parcel to Rice University, while retaining rights to the underground gas and oil reserves. The university gave 1,000 acres of land to NASA and also sold NASA an additional 350 acres for $1,000.00 per an acre.

With the close of the decade on the horizon NASA's astronauts as well as engineers, scientists and technicians had a job cut out for them. NASA flight director Gene Kranz acknowledged the momentous task: "We needed to race through our teens, and develop quickly." According Kranz's assessment the stunning Soviet Union's success served as a major incentive: "We were tired of being second."

Project Gemini would serve as an important bridge between Mercury and Apollo. Before mankind could go to the Moon several "firsts" must be completed, that would be the basis of the Gemini mission the mastery of steering, docking, maneuvering and un-docking (in preparation for the employment in NASA's Apollo Spacecraft and the lunar vehicle) and a study of the effect of prolonged periods of weightlessness for astronauts wellbeing (lunar missions were believed to last longer than a week, but NASA flight surgeons were still concerned about the consequences of zero gravity. Some feared that prolonged exposure to the weightlessness could cause death) Space walks to assess the efficiency of astronauts' suits for space (in readiness for walk on Moon) as well as the creation and use in fuel cell technology, not batteries that generate electricity and generate water (a essential requirement for the longer missions towards the Moon) and testing the maneuverability of the larger 2-man spacecraft (the Apollo missions would be expanded to three-man teams, with a bigger spacecraft) and testing the onboard flight and navigation systems and computers (the computer systems used by Mercury were housed in the control center for flights instead of within the spacecraft) as well as testing the more robust Titan II rocket that would be used to launch the spacecraft into the orbit.

Prior to the initial Gemini launch the Soviet Union's space program continued draw attention, causing many Americans to think that the country was getting further than it was with respect to this Space Race. On June 16th, 1963 Cosmonaut, Valentina V. Tereshkova, became first woman to fly in space. A few days earlier another cosmonaut Valeri Bykovsky had been launched into space and was awaiting Tereshkova's space flight. When both Soviet spacecrafts were orbiting in tandem, the spacecrafts moved close to 11 miles one another, which is a significant step towards the final goal of docking and rendezvous.

In the month of October 1964 Three Soviet Cosmonauts were launched into orbit on the Voskhad one spacecraft. To accommodate the very first three-man spacecraft The Soviets were obliged to take off all electrical devices inside a Vostok capsule. In the tiny quarters of the spacecraft the astronauts were not able to wear their heavy space suits that were pressurized, and they remained in orbit for just 24 hours.

Though it was short and in a makeshift manner however, the three-man Soviet mission was a first for space flight. Even with Gemini in the near future however, America United States still appeared to be catching up against Soviet Union. Soviet Union.

Assassination attempt on President Kennedy on the 22nd of November 1963, left NASA officials in a state of grief and grief. The man who had believed in getting to the Moon was abruptly and violently removed. Wernher Von Braun's secretary recalled JFK's death as the one and only occasion she saw her boss tear up.

In the wake of its sorrow, NASA redoubled its efforts to make a lunar landing by it was the time to end the decade. It was a living tribute to the fallen President. On Thanksgiving Day only six days following the death of his predecessor Lyndon Johnson announced that the Defense Department's Atlantic Missile Range and NASA's Florida Launch Operations Center would be changed to as the John F. Kennedy Space Center.

Two and a half months passed between the previous Mercury mission and initial Gemini mission. After two successful, unmanned flights, Gemini II, christened Molly Brown, manned by astronauts Gus Grissom and John Young began its mission on March 23rd of 1965. Grissom joked that he was going to name the spacecraft Titanic however NASA officials rejected the idea. The loss of Liberty Bell 7 during Project Mercury was too embarrassing an event to be remembered.

The Gemini spacecraft was successful and the astronauts travelled around the Earth 3 timesand spent 4 hours and 53 mins in space. While orbiting Earth Gemini II's crew Gemini II chartered new waters as well as was able to provoke controversy. With the help of this manual navigation device they were able maneuver the spacecraft between lower and higher orbits, a vital need for rendezvous and docking with a spacecraft. NASA experts and technicians were not amused when they discovered that Young had sneaked an uncooked corn beef sandwich into the spacecraft. When Young served the meal to Grissom during the spacecraft's orbit, pieces of the meat floated around the cabin, and were stuck to various instruments, earning both men harsh reprimands from NASA direction.

Five days before the launch of the first Gemini mission Gemini's first mission, the Soviet Union achieved another space exploration milestone in which Alexi Leonov made history as the first human to step into space. The tragedy was fortunately avoided by Leonov encountered difficulty entering Voskad 2. Voskad 2 spacecraft due to the difference in pressure between the spacecraft and capsule's air lock mechanism. This caused his suit of armor to erupt (akin to Michelin Man). A risky move, Leonov let go of the pressure in his suit. He only managed to crawl back inside the spacecraft until he was overwhelmed by exhaustion.

The problem continued to plague the Soviet crew after the spacecraft's auto-guidance system failed upon re-entry leading to the spacecraft landing at a distance of 2,000 miles outside the recovery zone. After landing parachute-style in a snowy forest the spacecraft's radio antenna and beacon were snuffed out by tree branches which made it much more challenging for searchers in finding the astronauts. Leonov and his co-pilot were forced to stay the night in the frigid spacecraft, which was surrounded with hungry timber wolves until they were found the next morning.

In the early part of June 1965, in the course of the Gemini IV space flight, Edward H. White became the first American to walk through space. Connected to the spacecraft through umbilical cables that provided life support, White used a nitrogen-powered zip gun to move in a zero-gravity environment, being orbited by the Earth at 18,000 miles an hour. When it was time to stop the spacewalk and return to in the capsule White declared his disappointment "It was the most depressing moment of my existence." White and his fellow crewmate, James McDivitt, attempted to connect with the Titan second stage rocket, but failed to connect with their falling orbital counterpart.

Onboard Gemini V, in August of 1965, Gordon Cooper and Pete Conrad spent nearly eight days in space and orbited Earth 120 times. They set an all-time record for the longest continuous space journey. As the astronauts continued tests of Gemini V's navigation and guidance system, they had to confront a problem in the fuel cell that had replaced batteries to provide the capsule's electricity source.

In the month of December 1965, Gemini VI, with crewmembers Wally Schirra and Tom Stafford and Gemini VII, manned by Frank Borman and James Lovell were both orbiting Earth simultaneously (another record) and sped within only a few inches of one another. Both teams were close enough that they could view each from their cockpits and proved that spacecraft can be aligned properly for docking. Gemini VII Gemini VII crew orbited the Earth for 206 times, and spent a record thirteen days orbiting the Earth.

In the month of March in 1966, in the March of 1966 aboard Gemini VIII, Neil Armstrong and David Scott completed the first successful orbital docking process with another spacecraft that was unmanned, the Air Force Agena upper stage rocket, which was 180 miles above surface of the Earth. The docking process was smooth until the spacecraft began spinning uncontrollably and roll. Armstrong was quick to eliminate Agena's Agena rocket, however it was the Gemini spacecraft continued to spin increasing to a speed of one revolution every second.

"We have a major issue there," Armstrong radioed Mission Control.

Disoriented and dizzy by the spinning spacecraft astronauts were on edge of losing consciousness at the time Armstrong chose to activate the engine system for re-entry.

"We all knew that if this didn't go as planned we'd be dead," Scott recalled.

The final second maneuver was successful and the spacecraft was able to recover after its demise. It was later discovered that the near-tragedy was the result of one of the thrusters of the spacecraft that was stuck in its in-place.

Gemini IX was supposed to be piloted by Elliot See and Charles Bassett However, both died by a plane crash on journey for the McDonnell Aircraft plant in St. Louis on February 28, 1966the first astronauts to be killed in the service. Gene Cernan and Tom Stafford eventually piloted the mission that launched on the 23rd of June the same year. The spacecraft was not able to join up with the Agena rocket after the docking mechanism failed in its ability to deploy fully. The spacecraft was saved, after Cernan completed long spacewalk.

Michael Collins became the first astronaut to perform a successful docking maneuver on an Agena rocket in Gemini X. Gemini X mission. The docking and rendezvous processes were replicated in Gemini XII and Gemini XI and XII flights.

Although it was obliterated by its Mercury predecessors as well as Apollo succeeding programs, the 10 piloted Gemini flights, which were conducted over a period of 20 months proved to be a vital bridge to American research into space. Mission Control Flight Director Gene Kranz summarized the legacy from the program's intermediary phase " Gemini developed the tools and techniques we needed for a trip to the Moon and, more importantly, Gemini was an important stage for crews and (flight) controllers." Neil Armstrong, who was the pilot of Gemini VIII, echoed Kranz's observation: "I believe that Gemini was timely and efficient. It gave many hours of experiences in the design for spacecraft."

As Project Gemini concluded, there were just four years left to go during the decade. Should the United States planned to reach the Moon prior to 1970 It could be sink-or-swim to launch Project Apollo.

## Chapter 5: Lucky Does Nothing To Do With Space Flight

In during the Mercury and Gemini times, during the Mercury and Gemini years, United States and the Soviet Union created lunar probes that were unmanned to gain a better understanding of Moon's orbit and topography. In the years 1961-65 NASA's Project Ranger launched nine such probes. After a string of failures The final three reached the Moon and then sent an estimated 17,000 high-detail photos of lunar surfaces back to Earth. With these images, NASA was able to find potential lunar landing spots.

Project Surveyor followed Ranger, with five of its seven probes navigating to the Moon between May until the month of July in 1968. The 2nd of June in 1966 Surveyor 1, a three-legged spacecraft, Surveyor 1 spacecraft landed on the Moon sending out 80,000 photos that provided NASA engineers and scientists with vital information about the lunar process of descent. After Surveyor 2 suddenly crashed into the Moon, Surveyor 3 successfully was launched in April 19th, 1967 which successfully sent 6,300 photos along with seismological and temperature data returning to Earth. Surveyor 3's robotic arm cut the lunar surface in order to determine the soil's composition.

In 1966 in 1966, it was in the year 1966 that the Soviet Union launched two unmanned probes (Luna 9 and Luna

10). Luna 9 was the first spacecraft to make an "soft landing" on the lunar surface. Luna 10 became the first spacecraft to orbit the Moon.

In the 1960s, the Cold War remained a virtual deadlock. The threat of nuclear destruction confined both the United States and Soviet Union to a stoic posture and prevented the onset from World War III.

In the midst of the Space Race, the two superpowers managed to reach some degree of peace. In 1967, the Treaty on Exploration and Use of Outer Space disclaimed any nation's claim to the Moon that would then be treated as international waters being the "property of humanity in all its forms."

As it was about to launch Project Apollo, the Soviet Union was in danger of being eliminated from its participation in the Space Race, a fact that was largely unknown public American public. When Soviet President Nikita Khrushchev was removed in 1965 and his successors failed to enjoy the enthusiasm for space exploration of the Sputnik period. The sudden demise of Sergei Korolev, the Soviet Union's top rocket engineer and designer of spacecraft in a routine, but erroneous surgical procedure was a significant setback to the nation's space exploration efforts.

The N-1 rocket which was the Soviets moon launcher proved to be temperamental. It required a volatile mix of kerosene and liquid oxygen for each stage to run The N-1 was prone to overheating. In four instances, N-1 test rockets exploded at their launch pads or just after the launch. Contrary to this The American Apollo Saturn V performed perfectly.

The fractious nature of bureaucratic conflict among Soviet scientist and the military's supervisors caused many delays in projects. The political ideology of Communism was one of conformity, but did not reward in a sufficient way the ingenuity of its people. Failure was usually followed with punishments, which included punishments like imprisonment, physical assault and/or execution. Likewise, successes were often rewarded with a plethora of propaganda-laden awards or medals. In such a chaotic system the scourge of paranoia was widespread and corruption was widespread, as the temptation to discredit colleagues in order to protect oneself (and possibly even one's life) could have hampered progress.

Contrary to NASA that was a non-military organization and was not a military agency, the Soviets made civilian engineers and scientists under the control of the military. A lot of Russian military leaders were more focused on weapons than space exploration and saw the possibility of the moon landing as unproductive and a waste of money.

The American lunar landing programme has been named after NASA project manager and engineer, Abe Silverstein: "I believe that the picture of god Apollo riding in his chariot through the Sun was the most accurate depiction of the magnitude of the programme." NASA Director James Webb described the broad objectives of the project: "The Apollo requirement was to depart from a spot at the top of Earth that was travelling at the speed of 1,000 miles an hour, as the Earth was rotating, and then to enter in orbit, at 18,000 miles an hour to orbit an object in space that was 240,000 miles away, is traveling at

2,000 miles an hour in relation to Earth in order to enter orbit around the body and then to bring a special landing vehicle onto the surface. The astronauts were required to take measurements and observations, gather specimens...and after that, repeat the backward-bound journey in order to return home...One of these missions could not accomplish the task. NASA needed to create an efficient system that could be capable of accomplishing this task time and time again." While simple as the goal seemed however, the preparation and execution of the lunar mission was laborious and exact, leaving little room for errors.

The Apollo mission to the Moon included a series of steps launch and orbiting Earth and docking the spacecraft to an lunar vehicle trans-lunar injection (escaping Earth's orbit and travelling toward the Moon) and lunar spacecraft orbits, descent of the lunar surface Moon walk along with lunar ascent, the docking process of lunar capsule and moonwalk the space capsule, the trans-Earth infusion (escaping lunar orbit before back to Earth) Re-entry into the Earth's atmosphere and recovery of the splashdown in the Pacific Ocean. Failure of the mechanical system or a pilot in any of these crucial steps could lead to the death of all three astronauts.

The development and design of the necessary components for the successful execution of the Apollo space mission took a collaborative effort that involved many thousands of engineers, designers and technicians both within NASA and in the private sector. The massive 3-stage Saturn V rocket, which cost 350 million dollars was the one responsible for launching the spacecraft into space. The concept was developed in the work of Werner von Braun, the Saturn V was constructed in an effort by North American Aviation, Boeing, McDonnell Aircraft, and IBM.

In the midst of pressure to get to the Moon at the close of the decade, NASA decided to launch the "all all-up approach" using Saturn V. Saturn V rocket. This means that all three stages of the rocket would be evaluated "live," on the initial flight, instead of separately.

It was the Saturn V was the first American space rocket developed and built for civilians only, with no military oversight. Due to its immense capacity, its massive first stage was not able to be fired with full force in the Marshall Center in Huntsville, because of the risk of breaking windows in homes nearby. Instead, the initial stage rocket was moved via barge across Gulf of Mexico and up the Pearl River to the Mississippi Test Facility (later named"the Stennis Test Center, in honour the Senator John Stennis). In this remote and swampy south Mississippi area, the rocket's powerful Saturn V engines were frequently testing fired.

On the 29th of January, 1964 it was the very inaugural Saturn rocket, a abbreviated two-stage missile launched out of Cape Canaveral. The rocket was able to launch into space the biggest payload of all time--a three-seven thousand pounds rocket launch stage.

The Command Service Module (CSM) which was designed to be used to transport Apollo teams from Earth to Moon in return, was designed in the United States by North American

Aviation. Lunar excursion modules (LM) was the brainchild from Langley Center aeronautical engineer, John C. Houbolt, was designed by Grumman. General Electric designed the fuel cells which power the spacecraft. Philco Aerospace Company equipped Mission Control with flight control consoles and IBM created Apollo's Apollo Computer Systems.

Politics and economics, the nursemaids and parents from NASA, the American space program eventually started to view their children as an unwelcome stepchild. At the end of the 1960s the U.S. Treasury was finding it difficult to fund its Vietnam War, Lyndon Johnson's famous and expansive Great Society social programs, and NASA's huge budget. Space exploration soon discovered that the cash cow of the government did not have an unlimited supply. In the decade that followed, the expenditures to explore space would be significantly diminished.

A tiny but vocal portion from the American public criticized space race Space Race as a misguided initiative. A growing counterculture of discontent mostly young Americans saw war-making and technological advancement as two giants that threatened peace and harmony, and they consumed minds and money which could be better used to tackle social problems, like poverty, hunger and illness. However, African Americans, the fastest growing part of the population (a direct consequence due to 1965's Voting Rights Act from 1965) considered that social issues that disproportionately affected minority groups, were being ignored by white males who were more focused on being on the Moon instead of helping the other men. Civil Rights Leader Whitney Young echoed the anti-Apollo opinions of his fellows: "A circus act--a marvelous trick that doesn't leave poverty unaffected. It will cost $35 billion dollars to send two people in orbit on the Moon. It would cost the sum of 10 billion to raise each one of us over the poverty line this year. There is something wrong with the system somewhere."

In the early months of 1967, preparations in preparation for the very first Apollo mission was well under way. Many believed that the successes from Gemini and Mercury Mercury as well as Gemini programs were the basis to a smooth transition however, others

warned against being too
confident. NASA Flight Director Chris Craft, reminded his colleagues: "We're making it look like it's easy. I'm hoping that we don't wind with a cost one day for leaving an impression that is false."

NASA was able to avoid fatal mishaps on the 16 space missions manned by astronauts that led the launch of Project Apollo, and none of the astronauts was seriously injured. The good fortune of the universe had smiled upon NASA's American Space Program, however Apollo Director of Flight, Gene Kranz, offered a concise message: "Luck has no business in space flights."

A sense of dread surrounded Apollo throughout its initial days. When the command service module was shipped in Cape Canaveral, NASA engineers expressed concern over the ship's "shoddy work." In the space capsule, the tangles of wires exposed caused concern for launch pad
engineers. Contrary to McDonnell Aircraft, manufacturers of the Gemini spacecraft North American Aviation and Grumman Aviation who built the command service module as well as the lunar module, did not divulge their

systems' details and schematics to NASA flight controllers. They were thereby not completing an essential element of the safety monitoring procedure. Astronaut Jim Lovell, summed up the displeasures of many of his coworkers: "The Apollo spacecraft, according to the most generous estimates, was proving as more of an Edsel." In the lead-up to the launch of the first Apollo flight in 1969, more than more than 20,000 system failures were documented. A sour Gus Grissom, who was scheduled to be the commander of Apollo 1, left a lemon in the flight simulator after having completed an exercise to train. Despite these worries, NASA moved headlong toward the first Apollo launch.

On the night of the 27th of January, 1967 Grissom along with his crewmates, Edward White and Roger Chafee were strapped into the cabin of their Apollo 1 command service module that was just over the Cape Canaveral launch pad. The astronauts were involved in a thorough "dress rehearsal" in preparation for the launch that was scheduled within three weeks. Due to delays due to mechanical issues that had occurred, the crew spent five hours in this cramped craft. Grissom who was who was a former veteran from Mercury and Gemini and Gemini, was aware of the dangers that space travel poses: "We flew with the conviction that if something truly occurred on the spacecraft, there wasn't any hope of rescuing." In all times, nobody was ready for a catastrophic event in a routine exercise.

at 6:31 p.m. In a flash with no warning Grissom shouted: "Hey!"

A few seconds later Roger Chafee called out: "Fire in the spacecraft!"

"Fire on the deck!" Edward White repeated.

NASA personnel from Cape Canaveral and Mission Control and Mission Control, who were monitoring this training exercise, were unable to believe their ears.

"We're burning! We need to get out of here!" Chaffee pleaded.

Television monitors displayed White trying unsuccessfully to open his cockpit's hatch. In just 20 seconds an explosion engulfed the crew as well as their corpses were scorched by the 2,500-degree (F) fires.

The culprit was found to be an untidy wire underneath the seats in the cockpit, which caused a spark in the oxygen-rich environment of 100 percent and ignited Velcro and papers for flight plans, foam padding made of polyester inside seats, as well as astronauts' combustible space suits made of nylon. They Apollo 1 crew had no chance of survival. To illustrate the astronaut's bleak situation and to illustrate the dangers of an 16.7-pound for every square inch of oxygen-rich environment, a lighted cigarette will be degraded in just two seconds. The heat created by the flash flame was strong sufficient to cause the melting of steel within this spacecraft.

The spacecraft's coolant ethylene glycol generated toxic fumes. Grissom was unsuccessfully trying to find to the lever which would've let out the oxygen-rich and noxious gasses outside of the spacecraft. Although it was equipped with the latest technology, the spacecraft didn't have one fire extinguisher.

Any chance to escape was destroyed because of the layout of the exit port. The outer hatch of the spacecraft was not able to be open until the internal one was taken out--a tedious processthat required the use of an torque wrench. In normal, non-emergent circumstances the crew and astronauts 90 minutes for the opening of the hatch. In a bizarre twist, Gus Grissom had argued against using an escape door that was explosive on the Apollo spacecraft, possibly tormented by his lingering embarrassment about that sinking incident of the Mercury capsule six years prior.

In a matter of seconds after the fire started starting to spread the capsule's pressurized chamber broke apart, fracturing its walls. In in vain, rescuers attempted to reach astronauts, but extreme heat and shock waves resulting from secondary explosions held the spacecraft from exploding. After launch pad crews finally got access to the spacecraft, a few of the rescuers burnt their hands while trying at opening the hatch.

When they got inside, NASA technicians witnessed unimaginable horror. Chafee was discovered strapped into his seat, completely charred beyond the point of recognition. Grissom White and Grissom White were found close to the hatch. Their melted spacesuits merged into the form of a groaning lump. Grissom (age 40), White (age 36) as well as Chaffee (age 31) were the first victims in the American space program. Grissom and Chaffee were buried in Arlington National Cemetery, while White was laid to rest at the school he attended, West Point.

Following after the Apollo 1 tragedy, all spacecrafts that were manned were shut down for 21 months, as NASA attempted to determine what went wrong. A committee of investigators within the NASA suggested 1,341 changes to the design of NASA's Apollo spacecraft.

Since it was simpler to utilize an all-gas supply to maintain cabin pressure 100 percent oxygen been regularly pumped into all American spacecraft. Following Apollo 1's tragedy, Apollo 1 tragedy, the capsule's atmosphere was changed to a mix of 60 % oxygen and 40% nitrogen to avoid spreading of the inferno. The use of flame-resistant Velcro paper, paper, as well as space suit materials were created, as were non-flammable cooling fluids. The hatch of the capsule was revamped to provide the crew with a rapid escape capability. If, in the future there was a fire in the capsule during training the astronauts would be given at the very least a chance to survive.

Following the Apollo 1 fire, NASA officials were devastated. some blamed themselves in the catastrophe. Apollo Flight Director Christopher Craft, was remarkably open about the incident: "We got in too in a god-damned hurry. We had to endure many of the worst equipment and inadequate preparation in order to finish the mission, and a lot of us knew that we were doing it."

A number of political leaders were becoming concerned about the huge budgetary strain caused by NASA and expressed their anger at NASA. In the congressional hearings relating to Apollo 1's tragic crash Apollo 1 tragedy, NASA Director, James Webb, drew a senator's ire: "The level of incompetence and recklessness we've witnessed here is simply unbelievable."

North American Aviation, builders of the Apollo spacecraft, was also in the responsibility. In 1972 the company paid the spouses of members of Apollo 1 crew a $650,000.00 settlement.

In honor of the memory that were the memory of Grissom, White, and Chafee, NASA cancelled the Apollo 1, 2, and 3 missions. It was the Apollo 4, 5, and six operations that followed unmanned tests.

Apollo 4 spacecraft Apollo 4 spacecraft was launched on the 9th of November 1997, the launch of the massive Saturn V rocket. The people who witnessed the launch weren't prepared for the earth-shaking noise of the enormous initial stage engine. The sound waves violently shaken on the CBS News television broadcast trailer situated three miles away from the launch pad 31 A. News anchor Walter Cronkite, placed his hands on the glass to prevent it from breaking while ceiling tiles fell: "My God, our building is shaking! The building is shaking! The roar is amazing! The big clear glass is shaking...Look at the rocket! go...Part of the roof has come in this area."

In just 11.5 minutes after the launch it was just 11.15 minutes since the Apollo 4 spacecraft was in orbit. Then, later in the mission the third stage of the Saturn rocket was launched and pushed the CSM to a higher altitude. On the 22nd of January in 1968 Apollo 5 was launched into space. On this test flight an unmanned lunar module was able to accompany with the CSM in orbit. After two successful flights and a successful launch, Apollo 6 was unable to achieve its mission. Apollo 6 spacecraft endured a number of issues during its orbit, which included issues related to the navigation system as well as 3rd stage rocket. Because of these problems the final Apollo mission that was not manned Apollo mission lasted just six hours.

After after the Apollo 1 tragedy, the general public's enthusiasm for exploring the moon began to wane. An Harris Poll, commissioned in July of 1967, showed that the majority of respondents opposed a lunar mission manned by humans with just 43 percent indicating their support. When asked what the Apollo project would cost 4 billion dollars annually only 34 percent said yes and 64 percent said that they did not.

On the 21st of October, 1968, 21 months following the devastating fire on the launch pad, Apollo 7 was launched. Astronauts Walter Schirra, Donn Eisle as well as Walter Cunningham spent 10 days in space in orbit around Earth sixteen times. The crew performed the first test in full on the CSM that performed very well. However, the mission wasn't without its share of controversy. Schirra was taken into space and was suffering from head colds, which was soon passed on to his fellow crew members. In zero-gravity, the astronauts' nasal passages wouldn't drain without blowing their noses, which was a challenge in space, even with space helmets. They also complained about poor quality of their freeze dried food stores, which turned into arguments about who had the right to most delicious food choices.

Schirra who had previously declared that it would be his last space mission was furious in his dealings with Mission Control, after the crew was directed to conduct not planned test: "I have had it all the way to today. From now on, I'm going to be on board as a Flight Director. ..." After the mission, Schirra expanded on his displeasures: "I had fun with Mercury. I enjoyed Gemini... Then I lost a friend and my next door neighbors, Gus (Grissom), one of our seven friends; I also lose two guys I imagined the world of. It was then that I realized it wasn't fun anymore. I was given a task that required me to bring it back to the same track as Humpty-Dumpty." After having decided to retire Schirra's candidness did not affect his professional career. Eisle and Cunningham did not have the same luck and neither of them flies into space for the second time.

It is believed that the Soviet Union endured its own terrible mishaps in the Space Race. On the 23rd of April in 1967 the Soyuz I spacecraft was launched into orbit and circled Earth for 18 rounds. When it reentered the retrorockets were fired too early, which caused the spacecraft to accelerate into the Earth's atmosphere at much greater speed than was intended. When the spacecraft's main as well as emergency parachutes became caught in the air, the spacecraft hit the ground at speeds of 400 miles per hour, killing the cosmonaut Vladimir Komarov.

In the month of March, 1968 Yuri Gagarin, the most well-known character from the Soviet space program, died while flying an MIG-15 aircraft trainer. The famous cosmonaut was honoured with his ashes being buried inside the Kremlin Wall. On the 8th of December 1968, an spacecraft that was not manned Zond 7 spacecraft, scheduled to orbit around the Moon was destroyed shortly after its launch at an altitude of about 27 miles.

Apollo 8, which launched on the 21st of December in 1968, was initially intended to evaluate both CSM and LM on the Earth's orbit. However, delays to construction on the lunar module prompted an alteration in the plans. Instead the astronauts Frank Borman, James Lovell along with William Anders became the first humans to depart the Earth's orbit and go towards the Moon. On the 24th of December in 1968, Apollo 8 entered a high lunar orbit of 69 miles and produced the first TV images of the lunar surface. The spacecraft stayed for a little longer than 24 hours orbiting in orbit, and circling the Moon many times.

The night of Christmas Eve, during its ninth orbit as the spacecraft came out of into the shadow of the Moon and television cameras filmed Earth at the horizon, the astronaut Bill Anders addressed a live television audience: "We are now approaching the lunar sunrise. For everyone on Earth Apollo 8's crew Apollo 8 has a message we'd like to share with you. At the beginning God made Heaven as well as Earth. The Earth was unformed and was void. Darkness was over the depths. And, the Spirit of God appeared on the face of the seas. Then, God said: 'Let there be illumination ...'"

In the following few minutes during the following few minutes, for the next few minutes, Apollo 8 crew took turns reading the first 10 verses from The Book of Genesis. They had ample reasons to call upon their faith. Prior to the launch, NASA experts calculated that the chances of the crew returning to earth alive were just 50. The astronauts ended their live broadcast by sending sweet greetings for the holiday season: "Merry Christmas and God bless you all--all of you here on this wonderful Earth."

A stunning photo of Earth captured during lunar orbit was published as the cover photo of Life magazine. It was reported that President Johnson was so impressed by the photo that he even sent the magazine to leaders around the world including the leader of North Vietnam, Ho Chi Minh.

Lunar mania soon spread. In the morning following Apollo 8's historic orbit Pan Am initiated its "First Moon Flights Club." For $14,000.00 members could secure seats on the proposed commercial Moon Shuttle, which Pan Am's founder, Juan Trippe, forecast to be operational by year 2000. Around 93,000 people joined the lunar exploration club comprising California Governor and the future president of the United States, Ronald Reagan.

On Christmas morning on Christmas morning, the Apollo 8 spacecraft blasted out of lunar orbit. This was a procedure called trans-Earth injection, which is the crucial initial step to return to earth. When the astronauts left for the Moon, Jim Lovell radioed Mission

Control: "Please be aware that there's Santa Claus." Santa Claus."

On the 27th of December, 1968 Apollo 8 splashed down in the Pacific Ocean, proving that human beings could make it a quarter-million kilometers into space, and then return safely. In the end, only a little doubt remained that the United States was clearly the leading contender in space race. Space Race.

Apollo 9, launched on March 3, 1969 was the first time that a test was conducted for the mechanical components that are required for the lunar landing. James A. McDivitt, David R. Scott, and Russell L. Schweickart orbited over 119 miles from Earth using the Command Service Module (CSM) known as Gumdrop. In lunar orbit, Schweickart removed from the moon module (LM) named Spider and separated it from the CSM and then was able to pilot it independently of the mothership. Schweickart along with McDivitt flew into Spider for six hours, flying more than 100 miles away from the CSM. Schweickart Scott and Scott also participated in spacewalks to test the suits of protection and backpacks with life support systems that could be used on excursions to lunar surfaces.

The complete rehearsal for the first Moon landing took place the 18th of May, 1969 the day that Apollo 10 blasted into space leaving Earth's orbit and headed toward the Moon. When it was in lunar orbit two astronauts Thomas P. Stafford and Eugene Cernan boarded the LM, Snoopy, and flew close to 50,000 feet the lunar surface. Stafford as well as Cernan were able capture the proposed lunar landing spot as well as provide NASA Geologists and engineers greater details about the topography of the region. As the lunar erupts, John Young remained alone in the CSM, Charlie Brown, and was able to orbit over his two crew members.

The Apollo missions 7 8, 9, and 10 were undamaged success, setting the stage for the ultimate prize: a lunar landing.

# Chapter 6: The Pioneers

Since the beginning people have gazed towards the stars of the night and thought about the bright spots of luminosity. Before the advent of cities there was a Moonless night sky has all the time been dark, and filled with stars that were so numerous, they couldn't be counted. Each of them was contained in the Milky Way galaxy, a continuous cloud that was home to thousands of stars.

William Gilbert (1544-1603), who realized the fact that Earth has two poles in magnetic field, conceived that other planets that orbit distant stars could have themselves magnetic fields, and maybe even life. In 1610 the famous Italian science researcher Galileo Galilei pointed his telescope towards Jupiter and found that it had four moons that revolved about it the exact way as the planets around the Sun. In relation to distance from their planet they comprise Io, Europa, Ganymede and Callisto.

It was not until nearly 45 years later when Dutch Astronomer Christiaan

Huygens spotted a massive Moon that was orbiting the planet Saturn. On the 25th of March, 1655 Huygens identified Titan which was the sole Moon within the Solar System with a thick atmosphere. The atmosphere has a pressure 1.45 times greater than the pressure at ocean level on Earth as well as Titan has a size of 5,150 km (1.48 times the size of the Moon) which makes Titan the second-largest moon in nature within the Solar System after Jupiter's Ganymede. Both Titan as well as Ganymede are bigger than the moon Mercury.

Huygens in trusting the memory of his self, saw holes of various sizes drilled into a screen that faced the Sun and determined the one that closely corresponds to its apparent luminosity Sirius which is the Dog Star that is the brightest star in the sky at night. Then, he calculated the vast distance between him and that star. If Sirius and the Sun and Sirius were intrinsically similar in brightness when seen at the same distance and scientists would've come close to his calculation. But, Sirius is an A1V hot, blue-white dwarf significantly brighter than Sun and is a G2V white dwarf with yellow hue. Sirius according to the accuracy of today is 22.3 times brighter than the Sun therefore, when they were placed side-by-side , at the 10-parsec distance (32.6 lighting years), Sirius would be 3.37 magnitudes brighter.

On the 13th of March, 1781, the German-born English scientist William Herschel discovered the planet Uranus. He initially believed that it was a comet but other astronomers determined its orbit and discovered it to be almost circular, similar to other planets.

In the year 1801, Italian priest Giuseppe Piazzi discovered a planet orbiting in the vicinity of Mars in 1801, and Jupiter. It was very small, with a radius of just 946 km (27 percent of the moon's diameter) in addition, Piazzi suggested that it be named "Cerere Ferdinandea," but the world-wide community of astronomers opted on Ceres. Presently, Ceres is referred to as"a "dwarf Planet."

The next calendar year German scientist Heinrich Wilhelm Olbers discovered a minor planet that was located between Mars and Jupiter that is now known as Pallas. The average diameter of the planet is 512 km, but the planet is far from completely spherical. Its dimensions are 476 km, 516 km and 516 kilometers. Due to its oval shape, it's not thought as a dwarf planet however, it is an Asteroid.

A few years after that in 1804 German scientist Karl Ludwig Harding discovered what is now known as the asteroid Juno having a median size of 247 km.

The year 1807 was when Olbers found his planetoid 2 in the belt of asteroid in the asteroid belt between Mars as well as Jupiter. Vesta has a size of 525 kilometers.

Out of all minor planets Between Mars and Jupiter of all the minor planets, there is only Ceres is thought to be a dwarf planet due to its size and shape that is near-spherical. There are currently thousands of planetoids that are believed to exist, with many of them between Mars and Jupiter however, others are in extremely elliptical orbits around the Sun.

It was in 1821 that French scientist Alexis Bouvard discovered some discordances in the orbit of Uranus when compared with the positions which had been calculated for it. Bouvard then proposed the idea of a distant planet in the vicinity of Uranus that was tugging at the same time on the planet that Herschel discovered. British scientist John Couch Adams estimated the parameters of the mysterious planet in 1843. Later, independently from Adams, French astronomer Urbain Le Verrier came up with himself the parameters of his suspect planet. Le Verrier sent his estimate to an German astronomer called Johann Gottfried Galle and asked to confirm the estimation. Galle discovered that the planet was located within 1 degree of arc from the expected position. Le Verrier named the planet Neptune that is recognized to have a median size of 49,244 kilometers (3.86 times the size of Earth). The collaboration of Le Verrier as well as Galle resulted in one of the more thrilling discoveries in science during the late 19th century. It proved that theory (the mathematical foundations of the celestial mechanical system) might lead to the discovery (the space itself).

Le Verrier

Galle

In the 19th century, researchers and scientists were thinking about stars and planets, considering whether any had life, and how humans could one day go there to discover. The early dreamers included Konstantin Tsiolkovsky, Jules Verne and Robert Goddard. Tsiolkovsky (1857-1935) was an Russian scientist who dreamed of stages for space stations and rocketry almost 100 years before they became feasible. Goddard (1882-1945) was an American pioneer in rocketry. He tested his theories of sending rockets in the sky further than space. To his credit, Verne created a new genre of science fiction that is now known as science fiction. his novels captivated the imaginations of millions of readers around the globe. The novel he wrote in 1865, From the Earth to the Moon was a novel about a journey to the satellite of Earth prior to the time that airliners were even allowed.

It's clear that the lack of the necessary technology did not stop people from dreaming of their own. Russian space scientist Konstantin Tsiolkovsky envisioned humans living in space, and tried to come up with ideas to enable it, such as airlocks, steering thrusters as well as space station.

In in the 1930s, 1920s American inventor Robert H. Goddard launched over three dozen rockets in the sky. Both his liquid-fuel rocket as well as multi-stage rockets were instrumental in paving the way to space flight. While neither of Goddard's rockets reached a height of more than two miles, his experiments showed that the technology could be used.

Goddard

It was in 1930 that Clyde W. Tombaugh discovered the dwarf planet Pluto which for over 70 years was considered to be an ordinary planet. In 2006, this changed when scientists revised their definition of planet and came up with the term "dwarf planet..

The science fiction writers in the late 20th century captivated young minds with tales of interstellar travel, even though the concept of flight was at its infancy. Televisions broadcast shows such as Rocky Jones Space Ranger and Flash Gordon, and movies began to portray interstellar travel in a more realistic manner such as that of 1956's

MGM movie Forbidden Planet. The film featured animated scenes created by an Disney artist who was loaned just to the film.

In the late 1950s America was adamant about rockets in the hopes of reaching space. And the technology that led to America's success in space can be traced to World War II, when Nazi Germany used rockets for bombing England. After the final rounds from World War II were fired and the reconstruction of Germany along with Europe began in the 1950s, Europe and Germany, Western Allies and the Soviet Union each attempted to secure the aid of the top Third Reich scientists, specifically those who were involved in missile technology, rocketry and research in aerospace. Naturally it was a sensitive matter because many members of these German researchers were just active Nazis and had also helped the Nazi war machine to terrorize the globe. In the same way during the war's final period and into the post-war period, those who were Western Allies formed a clear image of the Soviet state. Although they were forced to join with the Soviets however, the West began to realize that

Communist Russia represented a different aggressive totalitarian power and posed a serious risk towards an autonomous Europe.

The Western Allies and the Soviets were aware of Hitler's V-2 rocket project that was the precursor of ballistic missiles as well as the Space Race. Both recognized the strategic significance of these technologies, and both hoped to ensure their advantages for themselves. While the Soviets were contemplating expansion after the "Great Patriotic War" and the U.S. military came to realize that the supposed allies of today could become the enemies of the future, the people with a knowledge of V-2 rockets as well as the other Third Reich military technology programs were seen as vital pieces of the impending NATO to Warsaw Pact standoff.

It was the result of American led "Operation Paperclip" on the Western side and led to German experts putting their skills at the disposal of U.S. and other NATO members. Operation Paperclip aimed not only to gain the advantages of German technological advances to America but also to protect them from the United States but also to keep them from the hostile Soviets according to General Leslie Groves enunciated: "Heisenberg was among the most prominent physicists around the globe and during the moment of the German division He was more valuable to us than 10 divisions made up of Germans. If he fell into Russian control, he would have been a huge asset for them."

The Western method, though self-interested usually met with a consent on the part of German scientists' part. However the Soviet solution to Paperclip, Operation Osoaviakhim, employed the implicit threat of torture, imprisonment and even death, the most common tools used in Stalinist Russia to compel aid from German engineers and scientists after the conflict. They paid rich dividends to the Soviet state , in terms of attaining at minimum, a momentary technical level with western counterparts of the USSR.

To say that Operation Paperclip had a profound impact on the Cold War and American history is an understatement. The most famous example of this operation's "success" can be found in Wernher von Braun. Von Braun was part of a section of the SS who was involved in the Holocaust and would later be referred to for being"the "father for rocket technology" and captivate the world with his visions of flying rockets and space stations as an "new" Manhattan Project, one which NASA would later adopt. Alongside the development of ballistic missiles into weapons that developed throughout through the Cold War, von Braun's knowledge was utilized in the most significant space missions in American history. NASA was also required to design rockets that could launch the spacecraft into the Earth's orbit and later launch it towards the Moon. The Soviets struggled for years to create rockets capable of the task but with the help of von Braun, NASA got it right with its Saturn V rocket, which is still the most powerful rocket launcher NASA ever employed.

Von Braun

Following World War II, America established an organization called the National Advisory Committee for Aeronautics (NACA) however, after it was discovered that the Russians launch Sputnik in October 1957 the federal government was sufficiently concerned to approve the creation of National Aeronautics and Space Administration (NASA) in 1958. In the following years, President Kennedy would promise to send a man to the Moon before the close in the decade.

The first step on every space journey is to enter the orbit of Earth, which requires rockets that are powerful enough that a spacecraft can be launched, and accelerate it until it travels at a speed of nearly 18,000 miles an hour, which is the speed required for entering the orbit of Earth. The speed, precision and the technology required to reach orbit safely is difficult to attain even in the present, let alone more than a century ago and two of NASA's biggest catastrophes including the destruction from both the

Space Shuttle Challenger in 1986 and the loss of the Space Shuttle Columbia in 2003 were caused by problems that were encountered during launch.

When the spacecraft is placed in orbit and it eventually crashes back to Earth in the event that it fails to quit orbit. The speed required for an object to depart from orbit is called the escape velocity. To reach the escape velocity and escape Earth's orbit, spacecrafts need to be travelling at least 250,000 miles an hour.

While engineers were busy developing rockets that could get the object out of its velocity in the 60s, mathematicalists as well as scientists were still required to determine which planets were in their own orbits, and also when they would be there. As an example, as Mars as well as Earth have orbits that differ from their respective orbits around the Sun, Mars is in the same place in relation to Earth and the Sun at a single time in 780 consecutive days. This is called the synodic time which has to be calculated accurately. Spacecrafts within the Solar System will be orbiting the Sun throughout their travels therefore it's not as straightforward as shooting a spacecraft straight line, straight to the place where other planets are. Making it all work involves extremely complicated mathematic calculations. If the mission isn't completed at the right time, there might be a delay that could last for all of the synodic period.

The initial missions were dedicated to testing and building safe launch vehicles. Many mistakes - some rather spectacular made engineers and scientists discontent in the beginning. The solution to these early issues was not easy, but once they were resolved, it was possible to launch other missions. launched into space to accomplish different things, such as explore the the Earth's atmosphere. Explore space, observe the Earth in space explore the Moon and discover the characteristics of interplanetary space as well as the characteristics of the sun's wind and so much more.

In the mid 1950s, at the time that NASA was created scientists had no idea about the characteristics of space and whether human beings could live in the atmosphere of Earth. There was little information about the other planets of the Solar System, and what was learned came from hazy views of ground-based telescopes as well as through an unstable atmosphere on Earth.

When NASA created manned spacecraft, NASA was also working on several programs to gather the information needed to ensure an effective Moon landing. The beginnings of the Ranger were traced all the way prior to Soviet Union's rocket launch Sputnik in October of 1957. William H. Pickering, Director of the Jet Propulsion Laboratory in Pasadena, California, proposed that the United States respond by sending an spacecraft towards the Moon. Pickering was able to find a partner with Lee A. DuBridge, scientist and director of Caltech and both believing that lunar missions were the ideal American entry into the upcoming Space Race with the Soviet Union. R. Cargill Hall, historian at the Department of Defense, explained, "To his mind, and also to DuBridge's, lunar flights were a suitable entry point into the upcoming Soviet American space race. As rockets were a generation earlier lunar spacecrafts might were a subject just for science fiction however now they were the edge in engineering sciences, which was the place where

Pickering would have liked JPL to become. With the available technology and available, it was possible that the United States could launch a simple spin stabilized vehicle that was similar to the Explorer satellite at a very time-sensitive notice, perhaps by June 1958. Just three weeks before the launch Sputnik 1, Pickering, with the support of DuBridge, had prepared an JPL Moon flight proposal. Named 'Project red socks, the proposal declared it imperative for the nation to "regain its status in the global community through a major technological advancement against those of Soviet Union' in rocketry and space flight. Pickering demanded to get the Department of Defense to approve JPL's decision to launch immediately nine rocket launches to the moon. Moon."

DuBridge

Pickering

The proposal was not given an immediate approval from the Department of Defense, but in 1958, the newly established Advanced Research Projects Agency (ARPA) was able to take it into consideration as part of its mission to oversee the newly established U.S. space program. ARPA was receiving a variety of unadvertised proposals for projects and Vice Director Rear Admiral John E. Clark later said that "it appeared to me that everyone across the nation had come to the table with a proposal, except for Fanny Farmer Candy, and I was sure that they would come in to come in at any moment."

On the 27th of March in 1958, Secretary Defense Neil McElroy announced a lunar plan to coincide with this year's International Geophysical Year, consisting of three Air Force launches followed by two Army launches of a lunar probe developed by JPL. ARPA instructed to the Air Force to launch its probes "as as soon as is possible, in line to the condition that only a small amount of information that is useful about the Moon be collected."

The earliest Air Force probes designed by Space Technology Laboratories were described in depth in the following article by Hall: "Shaped like two cones that were truncated back-to-back the lunar probe made of fiberglass with a diameter of with a diameter of 74 centimeters (29 inches) in diameter and 46 centimeters (18 inches) in length, was able to carry 17.5 kg (39 pounds) of research instruments, batteries as well as an antenna and transmitter, as well as a retrorocket that was created to slow the spacecraft to a lunar orbit. As per the initial ARPA specifications, this spacecraft also had a tiny TV system that could be viewed as a facsimile. However, the engineers of Space Technology Laboratories Space Technology Laboratories had barely completed the spacecraft's design in June 1958, when the finding of the initial Van Allen radiation belt stimulated scientists to launch urgent requests for improved, more extensive experiments to study charged particles in the near-Earth space. Although they retained the television camera, the company's scientists focused the remainder of the instruments towards particles and fields in space, including magnetometers to study the magnetic fields that exist between the Earth and Moon and micrometeoroid impact counters to study the energy and flux of micrometeoroids that collide with these bodies. To gain more insight into what radiation distribution is within space, they set up an ion chamber supplied by

Van Allen on the second flight, and supplemented it with an inverse counter made by at the University of Chicago on the third flight. The report that was finalized summarized the scope of this massive modification: "To the greatest extent that is possible within the power and weight limitations the experiments were planned to provide scientific measurements of the cislunar environment.'"

The initial Pioneer mission was referred to as "Thor-Able 1" or simply "Pioneer." Its goal was to be able to orbit the Moon using a magnetometer micrometeorite detectors, and a TV camera. It was one of the very first attempts to launch the spacecraft outside the orbit of Earth.

Image of an Thor-Able rocket

The launch took place on the 17th of August 1958, the main aircraft (Thor) failed just 77 seconds into the mission, and due to the launch failed it was decided that the "Pioneer 1" designation was changed to "Pioneer 1. Engineers believed that the Thor booster's failure was caused by the turbopump bearing that became loose that would have stopped the flow of oxygen through liquids and caused a sudden decrease in thrust. The whole spacecraft and launch vehicle itself lost altitude and control, slanting to the left. The sudden change in direction is likely to have ruptured the oxygen liquid tank, which could be igniting with the fuel, leading to chaos.

The weight of the spacecraft's launch was only 83.8 pounds. It was believed to take 2.6 days to reach the Moon. When it arrived, the in-board rocket engine could have fired to slow the spacecraft down to settle in an orbit of around 18,000 miles over the moon's surface. If the mission was successful and scientists could have an idea of what the future astronauts could be expecting in the lunar region from the threat of micrometeorites. Also, it would have given images of the Moon much closer than anything before and also the first ever images of the Lunar Farside that had never been ever seen by a human. Humanity would be waiting until the 7th of October 1959, when the Soviet Luna 3 was launched. Luna 3 became the first spacecraft to photograph the previously unexplored part of the Moon.

Pioneer 0 was the first and the only Pioneer mission to be launched in the United States Air Force. The following Pioneer mission were launched under the brand new National Aeronautics and Space Administration (NASA).

It was the Pioneer 1 mission was also known as "Thor-Able 2" and again the goal was to travel around the Moon. It was a follow-up mission for the failed Pioneer 0.

It was launched on October 11, 1958 The Pioneer 1 mission had a easy liftoff, however the guidance system was malfunctioning and caused one incident following another, and ultimately resulting in inadequate momentum to exit the Earth's orbit. To keep the spacecraft alive in order to accomplish at least a valuable goal, NASA fired the onboard rocket motor. However, its trajectory wasn't precisely directed towards the higher Earth orbit it was seeking for the ad-hoc mission it was assigned. Instead, the spacecraft hit an altitude of 113,800 km (70,712 miles) before crashing back towards Earth. It was successful in analyzing the magnetic bands around the Earth, allowing scientists the ability to better understand what's now known as"the Van Allen radiation belt, however, technically speaking, Pioneer 1 "missed the Moon." Actually the spacecraft was not able to reach that point however, quick thinking enabled the newly created NASA organization to declare a victory in its very first Pioneer spacecraft mission.

A photograph from Pioneer 1 prior to launch

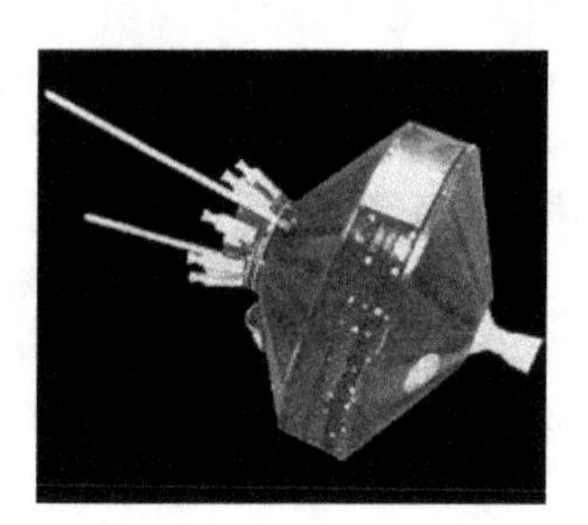

A photograph of the probe's structure

Pioneer 2 was also called "Thor-Able 3." and the mission was to be in orbit around the Moon and was the third attempt at doing the same thing that Pioneer 0 and Pioneer 1 did not succeed in doing. It was launched on November 8, 1958, the launch of the second and first stages proved successful, just like Pioneer 1, but like Pioneer 1, the third stage was not without its problems. The spacecraft reached only the height of 960 feet before being slammed back into the atmosphere , spanning Africa.

Pioneer 3 Pioneer 3 mission had a distinct goal from the three previous attempts. The spacecraft was only going to fly past the Moon and not try to achieve orbit around the Moon. With no necessity of a rocket motor The spacecraft was smaller and lighter, weighing 12.9 pounds, which is less than six times the weight of lunar orbiters from prior three mission. If all went as planned, Pioneer 3 would have been a fake Asteroid.

The launch took place on December 6, 1958, the liftoff from the launchpad was in line with schedule, however the first stage rocket motor came to a stop 3.7 second too late. Because of this, the spacecraft was unable to reach the speed required to reach the escape velocity. A number of other systems failed to work as planned, but ground control was prepared to maximize the chance despite the issues. Pioneer 3 was successful in reaching an altitude of 102,360 km which is about a third the distance to the Moon and then returning towards Earth, Pioneer 3 mission control specialists were able observe the spacecraft's measurements than the Van Allen radiation belt. The spacecraft was also able try out a trigger system that can trigger a camera that could be

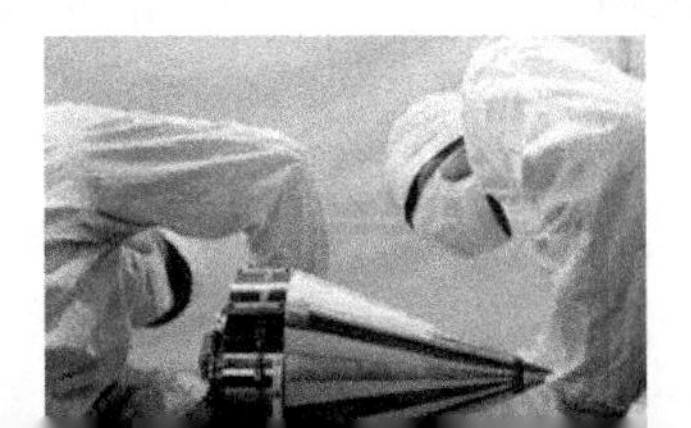

utilized on future missions.

A photo from The Pioneer 3 probe

Pioneer 4 was the only moon probe that was successful in the Pioneer program. This made it one of the very first American probe to escape the gravity of Earth and establish an orbit of the Sun. The aim was to examine the area within the Moon for radiation. A photocell in the tiny spacecraft was set to activate upon the time that Pioneer 4 passed to within 30 km of the Moon however, the closest distance to the Moon was 58,983 km (36,650 miles) and was not enough to trigger the sensor's photoelectric.

The launch took place on March 3, 1959 the rocket engines ran almost flawlessly, however the second stage of burns failed to finish at the right time, allowing the spacecraft an extra acceleration than it actually needed. However, the mission was a success, offering scientists with information on radiation conditions throughout the Moon and further. Telemetry data transmissions were in operation for 82.5 hours after the mission across the distance of 658,000 km (409,000 miles).

Its Pioneer 4 spacecraft achieved an orbit around the Sun using some of the parameters listed below:

Semi-major axis: 164,780,000 km.

Perihelion - 147,000,000 kilometers.

Aphelion - 169,000,000 kilometers.

It is a bit of a skeptic. 0.07109.

Inclination - 1.5 degrees.

Period The period is 398.0 day (slightly more than the Earth's calendar).

A photograph of the launch of Pioneer 4.

A version that is a replica Pioneer 4 probe

It was the Pioneer P-1 mission was also called "Atlas-Able 4A" or "Pioneer W." The mission of the spacecraft was to fly over the Moon using a TV camera, as well as magnetic sensors however, on the 24th of September 1959 during a pre-launch testing the rocket's motor exploded at the launchpad. Fortunately, the spacecraft was not yet installed in the launcher, and was later used in later in the Pioneer P-3 mission.

It was also known as the Pioneer P-3 mission was also named "Atlas-Able 4." "Atlas-Able 4B,"" in addition to "Pioneer AX." If the mission was successful and the mission was successful, this Pioneer

P-3 probe would have launched into orbit around Moon and utilized sensors to detect electromagnetic, micrometeorites and radiation fields electromagnetic radiation at various frequency and to establish an accurate moon's mass. The spacecraft also had its own rocket motors to make course correctionsand it was America's first test of the maneuverability of spacecrafts remotely.

The launch took place on the 26th of November 1959, the launch went without a hitch until the time of T+45 seconds. The pressure inside the spacecraft's shroud far exceeded that of the atmospheric pressure at that height. The result was an explosive decompression, which caused some of the payload shroud -- an element of fairing made of fiberglass which was to separate and separate from the launcher. Within the shield of protection, the support struts and spacecraft weren't aerodynamic enough to stand up to the chaos of shaking. The third stage rocket along with the payload was ripped off the launcher, putting an end to the mission.

This Pioneer P-30 mission was also named "Atlas-Able 5A" as well as "Pioneer the Y." Its purpose was to complete lunar orbit by using an array of instruments for science as a back-up mission to the unsuccessful Pioneer P-3. It was launched on the 25th of September 1960, the problem began when the fuel control were not shut down for the first stage, which caused the vehicle to keep burning until the first stage's fuel was exhausted. The second stage ignited correctly however thrust decreased to zero. The engineers later determined that the flow of the oxidizer been unable to function properly, and the fuel didn't have anything with to burn. The mission experts predicted that the spacecraft sank someplace in the Indian Ocean.

This Pioneer P-31 mission was also known as "Atlas-Able 5B,"" as well as "Pioneer Z." The mission's goal was to reach lunar orbit, as well as conduct tests similar to those designed for the earlier and failed Pioneer P-3 and P30 missions. It was launched on December 15th, 1960, the launch was smooth however, about sixty-six seconds in the spacecraft's mission the launch vehicle experienced rapid drop in the pressure of the oxygen tank in the liquid. The spacecraft crashed into the Atlantic Ocean roughly 16 kilometers away from Cape Canaveral.

As with all the unmanned space probes, every among these Pioneer lunar spacecrafts was able to fulfill particular objectives which would have provided concrete information to the space field Although these missions were mostly unsuccessful in technical terms they did have various kinds of success. One thing is that each mission was a testing of the launch vehicle's capabilities which would allow further space missions to be launched with greater confidence. The same way every failure was met by investigation that revealed little details that led to more efficient methods of launching the next spacecraft.

With the backdrop in the Cold War and the heating up of the Space Race, one of NASA's most difficult tasks was to market a spacecraft that seemed to be heading towards infinity. The Moon, a planet Moon as well as the Sun were easy targets to market, whereas deep space seemed to be too far literally and metaphorically. Naturally, NASA scientists realized they had to be aware of space in order to better understand the area around the planets as well as the Moon.

Pioneer 5 was also known as "Pioneer P-2,"" "Thor-Able 4" as well as "Pioneer V" which was the time when the goal was to study that interplanetary area that lies between Venus as well as Earth. It was clearly not as exciting as flying over the Moon or any of the planets, therefore initially, Pioneer 5 was supposed to fly over Venus but technical problems made it impossible to launch the mission in the Venus launch period in November 1959. After the technical issues were resolved in the mid-June 1960 Venus was moving away and it was not possible for Pioneer 5 to keep pace. A Venus mission had be awaiting the launch of its next window in which Venus as well as Earth were in their proper position relative to each other.

The launch took place on March 11, 1960. Pioneer 5 became the most successful of the entire sequence of "Pioneer-Able" missions. Launch itself proved fault-free, particularly in comparison to other Thor-Able launches. minor issues in the second stage were not sufficient to compromise the mission's overall success.

The weight of the spacecraft's launch is 43 kilograms (95 pounds). From the Spherical probe (0.66 millimeters in diameter) it was able to mount there were four "paddle-wheel" solar panels were extended to take advantage of sunlight for energy. The vehicle that launched the launch was an Thor DM-18 Able IV, departing at Cape Canaveral's LC-17A.

Following launch the spacecraft landed in an orbit that was heliocentric and elliptical between Earth as well as Venus and the following parameters for orbit:

Aphelion - 148,570,000 km (0.9931 the astronomical unit).

Perihelion - 105,630,000 kilometers (0.7061 AUs).

Eccentricity 0.1689. 0.1689.

Inclination - 3.35 degrees.

Period The period is 311.6 jours (more that 50 days less one year).

To give a comparability, here are the semi-major axis distances on Earth as well as Venus orbits:

Earth - the Earth - 149,598,023 (1.00000102 UAS).

Venus - the number of people who live there is 108,208,000 (0.723332 AUSs).

It is evident that Pioneer 5 traveled from slightly closer to the Sun than Earth to a place that was farther from that of the Sun that Venus.

With the help of Pioneer 5, scientists learned that interplanetary space actually does possess an electromagnetic field.

Pioneer 5 was a ship carrying four important scientific instruments:

Solar wind detector - used to detect solar particles, which includes electrons that have an energy higher than 13. MeV.

CPSIA information can be obtained
at www.ICGtesting.com
Printed in the USA
BVHW050134070223
657977BV00015B/719